THE SCIENCE OF STAR WARS

This book is unofficial and unauthorized. It is not authorized, approved, licensed, endorsed or connected in any way with Lucasfilm, Disney, or Hasbro.

THE SCIENCE OF STAR WARS

THE SCIENTIFIC FACTS BEHIND THE FORCE, SPACE TRAVEL, AND MORE!

MARK BRAKE AND JON CHASE

Racehorse Publishing

Racehorse Publishing books may be purchased in bulk at special discounts for sales promotion, corporate gifts, fund-raising, or educational purposes. Special editions can also be created to specifications. For details, contact the Special Sales Department, Skyhorse Publishing, 307 West 36th Street, 11th Floor, New York, NY 10018 or info@skyhorsepublishing.com.

Racehorse Publishing™ is a pending trademark of Skyhorse Publishing, Inc.®, a Delaware corporation.

Visit our website at www.skyhorsepublishing.com.

10 9 8 7 6 5 4 3 2 1

Library of Congress Cataloging-in-Publication Data is available on file.

Cover design by Brian Peterson
Cover photographs: iStockphoto

Print ISBN: 978-1-944686-28-4
Ebook ISBN: 978-1-944686-29-1

Printed in the United States of America

TABLE OF CONTENTS

Part II – Space

Part III – Aliens

Part IV - Tech

Part V - Bio-Tech

INTRODUCTION

I have a long history with Star Wars and science fiction.

In the same summer that brought *Star Wars: Episode I – The Phantom Menace*, I designed and validated planet Earth's first science and science fiction university degree program.

The world's press had a field day. The degree program would enable its students to study "science fiction—from the publication of Mary Shelley's *Frankenstein* through to the current Star Wars movie." Back in 1999, that first syllabus examined "the link between science fiction and 'real' science," and considered the "cultural significance of science and science fiction." For the better part of two decades since, I've explored the relationship between space, science, and culture.

This book does just that.

Through the lens of Star Wars, we see the universe in a new light. Almost on a daily basis science unveils something new, and potentially shocking, about the world in which we live as well as its relation to the rest of the universe. The countless galaxies wheeling through space and time. The rise of the robot. The unraveling of the human genome. The universe, it seems, is increasingly decentralized, infinite, and alien. In short, the universe is a strange place.

Star Wars is a response to this cultural shock—the shock of finding ourselves in an increasingly marginal position in a hostile cosmos. The stories of Star Wars help us come to terms with this new universe unveiled by science. Star Wars works by conveying the taste, the feel, and the human meaning of the findings of science. Star Wars puts the stamp of humanity back onto the universe. It makes human what was once alien.

Sure, maybe there *are* more stars in the universe than grains of sand on all of Earth's beaches. Perhaps a hundred billion galaxies flow in the great beyond, outside our infinitesimal Milky Way. But Star Wars makes you feel we could still *own* our Galaxy. Earth may no longer sit at the center of the universe. The Sun may no longer be the only star with planets. But with small steps and an outward urge (and preferably our own *Millennium Falcon*), that galaxy could someday still be ours.

If the science of space isn't daunting enough, there's Darwin. Man among the microbes, with no special immunity from the jaws of evolution, and vanishingly little evidence of a divine image. Each successive biological discovery to date has had a huge impact, both on the human condition and on the meaning of life in the universe.

Star Wars helps.

It helps us to imagine how evolution might play out on a cosmic scale. Darwin's theory of evolution cuts both ways. It provided a mechanism for the evolution of life on Earth. It also delivered the self-same mechanism for how alien life might develop and evolve in extraterrestrial settings. Evolutionary biologists may conjure up all sorts of scenarios. But Star Wars puts those scenarios into sci-fi, and the theories of science come back to us transformed by fiction.

That's why there is something revolutionary about Star Wars. Painting pictures of the cosmos, as it were, has a dialectic effect—the cosmos comes back to us changed. By imagining the strange worlds of Star Wars, we come to see life in the universe in a new perspective.

It's so easy to be dazzled by the infinity of visions Star Wars presents—a bewildering diversity of star systems and planets, alien species and spaceships, droids and cyborgs, light tricks and life forces.

On a simpler level, Star Wars is all about the relationship between the human and the nonhuman. So that's how this book is structured. It's divided into five conceptual themes: space, space travel, tech, bio-tech, and aliens. Each of these themes is a way of exploring the relationship between the human and the nonhuman. Taking a closer look at these themes will illuminate the genius of Star Wars, as if a vibrant lightsaber were being held to it. It will show the way in which the franchise functions.

Space. Space in Star Wars is a vast arena in which the stories unfold. But it's also a facet of the nonhuman, natural world, replete with stars and habitable planets.

Space Travel. Having the vast arena of space is one thing. But how do you get from one star system to another? This theme deals with nonhuman questions of journeying to the stars, including faster-than-light travel, hyperspace, and a certain gargantuan space station.

Tech. What will the future bring in the form of machines? Star Wars has plenty to say about the rise of the robot, and the prospect of artificial intelligence. And what of the tech of social engineering and the surveillance culture of the Empire?

Bio-tech. What might humans one day become? Whether through genetic design or bio-tech enhancement, Star Wars peeks at our wetware evolutionary future.

Aliens. If space truly is a vast arena of habitable planets, what kind of creatures lurk in its depths? Star Wars has created some of the most famous alien life-forms in all of film and fiction.

So join us in our open-minded search for science in the world's most popular science fiction franchise.

Mark Brake, 2016

PART I
SPACE TRAVEL

HOW HAS STAR WARS INFLUENCED SPACE CULTURE AND SPACE TRAVEL?

In March 1944, a detail of US Army counter-intelligence agents raided the offices of the monthly magazine, *Astounding Science Fiction*. Their brief was simple: to uncover any data leaks following the publication of a speculative sci-fi story about an atomic weapon. This was decades before the Death Star, but bombs were all the rage even then.

The spooks didn't do a very thorough job.

After the raid, the magazine's editor, John W. Campbell, voiced his relief that the agents failed to spot a wall-map in the office, which detailed the distribution of subscribers across America. Clearly marked on the map with bright-red pins was a cluster at PO Box 1663, Santa Fe, New Mexico. The address was home to the Manhattan Project, the Allied attempt to build the atomic bomb, which, like the Death Star, was another super-weapon invented in science fiction.

But the plot was thicker still. Consternation at Counter-Intelligence HQ would have reached fever pitch had they known that one bright-red pin told another tale. For the duration of the war, a copy of *Astounding Science Fiction* was regularly imported into Germany by one Wernher von Braun, a member of the Nazi party and the SS, and chief scientist in Hitler's quest to build a Nazi atomic bomb.

The event is a crucial turning point in science fiction mythology. And it is a prime example of the way in which sci-fi, especially Star Wars, has come to possess a dominating influence on contemporary culture.

Science Fiction is No Subculture

Star Wars is a major inspiration for a space-obsessed culture that makes us the first generation to live in a science fiction world.

Media headlines trumpet the discovery of exoplanets that bear more than a passing resemblance to Tatooine and Hoth. As we switch on the TV, we see the prospect of driverless droids, orbiting space stations, and interplanetary robots that rendezvous with asteroids. Scientists tell us that the spacecraft of the future will use solar sails in a similar way to the solar panels used by the Empire's TIE starfighters.

In China in 2007, they held the first party-approved science fiction convention in Chinese history. To quote a lecture given by Neil Gaiman:

> *"[Science fiction] had been disapproved of for a long time. At one point I took a top official aside and asked him what had changed? 'It's simple,' he told me. 'The Chinese were brilliant at making things if other people brought them the plans. But they did not innovate and they did not invent. They did not imagine. So they sent a delegation to the US, to Apple, to Microsoft, to Google, and they asked the people there who were inventing the future about themselves.'"*

They found a common link: all had read science fiction when they were young, and Star Wars had inspired many.

War in Space, Anyone?

The most dramatic early impact of such inspiration was the rise of the "Star Wars" Strategic Defense Initiative (SDI).

Ironically, between the launch of *The Empire Strikes Back* and *Return of the Jedi*, NASA's space science program had been all but scrapped. Most future plans for the peaceful exploration of space had been abandoned. Someone was taking the idea of war in space very seriously, as military space missions were mushrooming. In 1981 alone, the non-classified budget for such military projects was over ten billion dollars, with further untold billions in covert missions. H. Bruce Franklin's *War Stars* reports that by the end of the war the civilian director of the shuttle mission had been replaced by a chief architect of space war. Thus, SDI was born.

SDI had a backstory.

The launch of *Sputnik* in October 1957 had not just launched the Space Age, fired off the space race, and heightened the Cold War. It had also sent shock waves across America.

Future President Lyndon Johnson declared in a speech only months later: "Control of space means control of the world. . . . There is something more than the ultimate weapon. That is the ultimate position—the position of total control over Earth that lies somewhere in outer space." The Manchester *Guardian* in the UK focused sharply on underlying American fears: "The Russians can now build ballistic missiles capable of hitting any chosen target, anywhere in the world." American paranoia peaked when it was realized that *Sputnik* had flown over the US not once but four times during its brief voyage.

"Star Wars" (SDI)

And so was born the idea of attack from space.

The Soviet rocket that launched the first human-made object into space also brought home to Americans the threat that had already hung over the Soviets since Hiroshima. A rocket that could put a satellite into orbit could just as easily send a superweapon on a ballistic trajectory to the United States.

By the early 1980s, the Star Wars franchise had crystallized the paranoia into SDI.

SDI was a hugely ambitious superweapon that aimed to protect the US from attack by ballistic strategic nuclear weapons. The SDI missile defense system combined ground-based units with a space weapon, sometimes known as orbital deployment platforms. This is where much of the Star Wars influence comes in. SDI would attack targets from space, or disable missiles traveling *through* space.

Someone really had taken the Death Star to heart.

The space-weapon part of SDI ranged over a number of concepts, but they included groups of interceptors, which were to be housed in orbital modules, and the idea of orbiting satellites carrying powerful lasers or particle beams. There we see the direct similarity with the Death Star. In 1987, the American Physical Society said that a global shield such as SDI was extremely ambitious and that existing technology was not directly feasible for its operational status. In other words, it wouldn't work. The program was eventually abandoned. After thirty billion dollars spent, no such space weapon system was ever used.

Fact Blurs with Fiction

NASA survived the 1980s infatuation with military space missions.

Instead, NASA has now boldly jumped on the Star Wars bandwagon. Gone are the sober Cold War mission monikers of *Apollo*, *Mercury*, and *Magellan*. Gone are the grandiose references to hifalutin Greek and Roman gods. Realizing that many of its scientists and engineers take much of their creative inspiration from the franchise, NASA turns to Star Wars to seduce an increasingly media-savvy generation. Consider the illuminating words of NASA astronaut Kjell N. Lindgren, flight surgeon and engineer aboard Expedition 44/45 in July 2015, on the International Space Station:

> *"Star Wars is definitely the first movie I remember seeing. I must have been three or four at the time. I'm also just a big fan of science fiction and speculative fiction in general. And my father was in the Air Force, and I grew up on Air Force bases. I think all those things taken together influenced my desire to become an astronaut. It's what I wanted to do for as long as I can remember. I think Star Wars, just the story, of course, captures the imagination. And then also the technology and the idea of living in space and doing all those things is very exciting as well. I've never let go of that. It's interesting because Star Wars is a cultural touchstone, and we're*

in a generation of astronauts now that saw A New Hope, The Empire Strikes Back, *and* Return of the Jedi *when they were little kids. Star Wars is one of the many reasons that I became interested in space flight."*

Kjell Lindgren also conjured up the notion of a Star Wars-themed mission poster. NASA gave the idea its backing. The poster shows the crew kitted out as Jedi Knights, lightsabers in hand. Behind them we spy a starfield, with an X-wing type satellite. And the official name of the mission is "Space Station: Expedition XLV–The Science Continues!"

Lindgren is not alone. NASA booths are a frequent sight at Star Wars conventions. The agency engages in Star Wars "May the Fourth" Day celebrations. NASA has a commercial crew and cargo program named C-3PO, which wishes to "extend human presence in space by enabling an expanding . . . space transportation industry." And it was NASA who dubbed Kepler 16b "Tatooine," saying "the existence of a world with a double sunset, as portrayed in the film Star Wars more than 30 years ago, is now scientific fact."

Star Wars and Science

Science fiction like Star Wars first emerged with science. Way back at the time of the scientific revolution, Earth became an alien planet. When scientists made the earth-shattering suggestion that we did not live at the center of the universe, the revolution cut two ways. It made earths of the planets, and it also brought the alien to Earth. The universe of our ancestors had been small, static, and Earth-centered. Humanity was its guiding light. The new universe was decentralized, inhuman, and dark.

So, Star Wars stories of space voyages help us make sense of the nonhuman universe in which we find ourselves: our marginal position in space, our fate in time, the rise of the robot, and the unraveling of the Sith monster within us. Star Wars is a way of exploring the relationship between the human and the nonhuman aspects of the universe, as unveiled by science. The franchise allows us a fictional look at how the science of the future might pan out.

Star Wars operates in the way that science fiction always has. It sometimes takes a poet, or a movie director, to best express the taste, the feel, and the human meaning of scientific discoveries. It's a way of describing the cultural shock of discovering our marginal position in alien space; an attempt to put the stamp of humanity back onto the universe, to make human what is alien.

In this way, it is easy to see that Star Wars is a special way of thinking about science. It speculates on the new worlds uncovered by discovery and exploration. It uses the fantastic, strange worlds of the imagination to come to terms with our conditions of life in a new and potentially revolutionary perspective.

The franchise has shaped the way we see and do things, the way we dreamt of things to come. It helped us discover the familiar in relation to the unfamiliar, the ordinary in relation to the extraordinary, and forced us to explore the nature and limits of our own reality.

Star Wars helped us build the future we now live in.

COULD SPACERS TRAVEL TO OTHER GALAXIES IN THE STAR WARS UNIVERSE?

Consider a tale of two universes. The point of this thought experiment will be how a space-farer might travel from one of these universes to the other.

The first universe, the factual universe, is the one in which we live. The *other* universe is the fictional Star Wars expanded universe. This universe incorporates all officially licensed Star Wars media material, including books, comics, video games, toys, TV movies, and the whole host of franchise paraphernalia, whereas the media created by Lucasfilm is considered official Star Wars canon: the Star Wars films, the Clone Wars TV series, and the Rebels series.

The Walt Disney Company acquired Lucasfilm in November 2012. On April 25, 2014, Disney announced that all previously released expanded universe media would be rebranded as Star Wars Legends.

Star Wars canon, and Star Wars Legends.

A new Disney division, Lucasfilm Story Group, would in future warrant that all forthcoming media were non-contradictory to the films, each other, and works written since the announcement. Given they are part of a similar story, however, content and characters based on Legends media may appear in the new "storytelling approach." On June 16, 2016, it was announced that Legends content would be included as downloadable content in the video game *Star Wars: Battlefront*.

It's a very fluid fictional universe.

The Factual Universe

For most of the time in this book, we've stayed on the canon side of things.

As can be seen from the introduction above, the fictional waters of the Star Wars expanded universe can be very murky indeed. It's little wonder that folk talk about dark matter, and the dark side. Plot lines may diverge. Characters may exist in one reality and not in another. More importantly, for our purposes of exploring the science in Star Wars, concepts claimed about the universe at large may be present in Legends, but not in canon.

And yet, whether canon or Legends, that galaxy far, far away simply must be situated in this *factual* universe, the one in which we live, and which shares a similar science and philosophy with our own Galaxy. We might even call this universe (the one shared between the Star Wars galaxy and our own Milky Way) an expanded *factual* universe. At least, that's the assumption we've made throughout this book, and it seems like a reasonable one.

But what does Star Wars, whether canon or Legends, have to declare about the wider universe beyond that fictional galaxy? And does the declaration bear any resemblance to the factual universe?

Beyond That Galaxy Far, Far Away

According to Legends, the Star Wars galaxy had close companions.

In terms of intergalactic travel, these satellites would have been very doable destinations. There were seven satellite dwarf galaxies in orbit around the galaxy proper. They were named, in alphabetical order: Aurek, Besh, Cresh (I'm not making this up), Dorn, Esk, Forn, and Grek. Apart from the names, all of this is perfectly in order. Our Milky Way Galaxy also has satellite galaxies in attendance. About fifty of them, in fact. The only satellite galaxies visible to the naked eye are the Large and Small Magellanic Clouds, which, even though they've been observed since early civilization, were nonetheless named for a Portuguese explorer who rather belatedly spotted them in 1519.

Of the Milky Way satellite galaxies confirmed to be in orbit, the largest is the Sagittarius Dwarf Elliptical galaxy. The Sagittarius Dwarf has a diameter of 8,500 light-years, which is almost a tenth that of the Milky Way's. Curiously, given its size, Sagittarius Dwarf wasn't discovered until 1994. Even though it's one of the closest companion galaxies to the Milky Way, it's on the opposite side of our galactic core from the Earth. As a result of its sheer size, even though it's very faint, it does cover quite a large area of the sky in that direction.

Intergalactic Travel

Does Star Wars admit to its own satellite galaxies as doable destinations?

It does indeed, at least in Legends. Legends admits to the main Star Wars galaxy as having been home to between five and twenty million different sentient species. Over one hundred quadrillion aliens are said to have lived in one billion star systems. These species interacted with one another through diplomacy, through trade, through war, and by necessity, through travel.

Of the satellite galaxies, companion Aurek was also known as the Rishi Maze. This was a reference to the planet Rishi, located within Aurek, which was the starting location for hyperspace travel into the companion galaxy itself. Some of the companions were said, again in Legends, to house some twenty billion stars. The companions were ranked in order of distance, with companion Besh being the furthest. Also known as Firefist, Besh is some 150,000 light-years away from the main galaxy, and is only ever surveyed by probots—probe droids specifically designed for recon missions.

Is this ringing any factual bells? Pretty much.

Of the fifty or so satellite galaxies of the Milky Way, around eight are within 100,000 light-years. Of those, the Ursa Major II Dwarf (Uma II) galaxy sits almost at the 100,000 light-year boundary from the Milky Way. Uma II has a mass of about five million Suns, and is thought to have formed at least ten billion years ago. The stars of Uma II were probably among the first to form in the universe.

That Good Old Hyperspace Disturbance

So Star Wars admits to intergalactic travel, but it's limited.

Star Wars spacers could journey to the satellite galaxies, but no further. Why? The small matter of a hyperspace disturbance beyond the edge of the galaxy. This "disturbance" seriously complicated extra-galactic exploration, as a band of whorls and eddies spun around the galaxy's edge too swiftly to be crossed by faster-than-light (FTL) speeds. Beyond the galaxy's rim was a vast expanse of starless space, known as the Intergalactic Void.

Now, to those of a cynical turn of mind, this sounds very much like a "MacGuffin." The idea of a MacGuffin was popularized by British movie director Alfred Hitchcock, and it describes a plot device in the form of a goal, desired object, or, in the case of a "hyperspace disturbance beyond the edge of the galaxy," an obstacle that drives the narrative forward. Or rather, *doesn't* drive the narrative forward. The writers of Star Wars Legends may have felt that, with one billion star systems and between five and twenty million different sentient species, they already had enough to explore. Their narrative was already sufficiently driven.

Both film and fiction have often used the concept of a remote and "alien" island as the venue for a narrative. An island setting places the action beyond the "real" world, presenting an image of an ideal, such as a utopia or dystopia. This ploy goes back at least to Thomas More's *Utopia*, and includes other famous works such as *Robinson Crusoe* and, of course, *Jurassic Park*. In many ways, it even applies to Harry Potter film and fiction, as to a large extent the narrative unfolds at Hogwarts School of Witchcraft and Wizardry, a remote "island" setting beyond the real "muggle" world of non-magic.

The description is even more fitting for the Star Wars galaxy when you realize that, in the very early days of galactic exploration, German philosopher Immanuel Kant described the distant galactic nebulae, spied through the giant telescopes of the time, as "island universes."

Beyond the Hyperspace Disturbance

Sky surveys and cosmic mapping of the factual universe have been carried out on a gargantuan scale.

What have the mapmakers learned about cosmic architecture? The large-scale structure of our universe begins at the star level. Stars are organized into galaxies. Little new has been discovered about that. However, beyond that the cosmic structure is a continuum. Scientists believe there are a number of levels of structure larger than galaxies—groups, clusters, super-clusters, walls and filaments, many millions of light-years across. The walls and filaments are sometimes separated by immense voids of a similar size, creating a vast, sponge-like structure that is often called the "cosmic web."

Therefore, Star Wars is right. There *is* an intergalactic void, or maybe more like an interfilament void. And it *is* huge. However, the void wouldn't stop spacers traveling from galaxy to galaxy, through groups and clusters, along filaments many millions of light-years long.

HOW MUCH WOULD IT COST TO BUILD A DEATH STAR?

Just think about the sheer size of the Death Star.

Death Star I, or *DS-1 Orbital Battle Station,* as it's also known, was *moon*-sized. In our solar system, moons come in all sizes. Some of the sixty-odd designated moons in orbit around Jupiter are tiny, only one or two miles in diameter. Saturn's moon Titan, the biggest of the planetary moons in orbit about the Sun, is a huge 3,200 miles across. Our Moon is 2,158 miles in diameter, about a quarter of the size of Earth.

The diameter of *Death Star I* was seventy-five miles.

So, that's a moon-sized (and shaped) space station, with its own gravitational pull, and enough living space inside for a crew of 342,953 Imperial Army and Navy, and a further 25,984 stormtroopers. That's some construction.

Add to that the kind of creature comforts most other Imperial military postings simply didn't have: recreation areas, cantinas with state-of-the-art bartender droids, and, famously, a canteen, and you begin to get a better idea of the scale and cost of this venture.

The Build

Let's build one. Firstly, steel. Lots of it. In fact, gargantuan amounts of the stuff. Assuming around one tenth of *Death Star I* was something other than empty space, the build would need about 134 quadrillion tons of steel. That's 134 thousand trillion, or 134,000,000,000,000,000 tons.

That amount of steel would cost the Imperial Treasurer around $852 quadrillion ($852,000,000,000,000,000). And cost is not the only challenge. Consider the production rates. At the frequency the entire Earth currently produces steel, *Death Star I* would use up around eight hundred millennia's worth. Yes, it would take us humans 800,000 years to make that much steel!

Death Star I was constructed above the inhospitable, desert world of Geonosis. This raises another question. As Geonosis alone is very unlikely to yield enough steel for Death Star production, resources would have to be shipped in from other worlds, at great expense. Shipping into space, off any productive planet, would cost an approximate one hundred million dollars per ton.

However, there is a second option on the *Death Star I* steel build: asteroid mining.

Here on Earth, asteroid mining is space's new frontier. A class of asteroids known as easily recoverable objects (EROs) was identified by researchers in 2013. Twelve asteroids made up the initially identified group, all of which could be potentially mined with existing rocket science. Of nine thousand asteroids searched in the Near Earth Orbit (NEO) database, these twelve could all be brought into an Earth-accessible orbit relatively cheaply, and by changing their velocity to less than 1,100 miles per hour.

Furthermore, consider the solar system asteroid 16 Psyche. It's believed to contain a small portion of precious metals, but more importantly, 1.7×10^{19} kilograms of nickel–iron. That's enough to supply Earth's production requirement for several million years. And 1.7×10^{19} kilograms of nickel–iron is equivalent to 17,000,000,000,000,000 tons of the 134,000,000,000,000,000 tons needed to build *Death Star I*. Just another seven asteroids like 16 Psyche and we have enough metal for the build.

To save further on to-and-fro shipping and building costs, the necessary asteroids could be transported to a safe orbit in the vicinity of the Death Star construction. This could hypothetically allow for more materials to be used and not wasted. However, compared to the overall build costs, the savings would be nominal.

The Cost

The first Death Star was built using Wookiee slave labor and various other alien species to complete the project. Both during and after construction, organic lifeforms would have needed air to breathe.

Assuming six-tenths of *Death Star I* was pressurized space, the project would require 8.23 quintillion cubic meters of nitrogen (the main constituent of air) and 1.65 quintillion cubic meters of oxygen. Delivery would set you back $3.48 septillion and $263.33 quintillion, respectively.

Once it's all been added up, the total cost of *Death Star I* comes to a roaring twenty septillion dollars, or $20,000,000,000,000,000,000,000,000. That's about one trillion times the current US debt, or the cost of two thousand trillion missions to Mars, which would probably be enough to transport the entire population of Earth to Mars.

And that's only your bargain basement Death Star.

We haven't yet added crew quarters, life support systems, computer/AI networks, wifi, power generators, 5-star luxury options, aircon, mega-lasers, Darth Vader's penthouse, or the canteen. Or construction costs. We all know how builders' estimates turn out to be fiction rather than fact. As for *Death Star II*, it was one hundred miles across and housed 560 internal levels to accommodate approximately two and half million passengers. The build cost of which is left as a casual exercise for the reader.

Down With the Death Stars

But here's the dark side lurking in the detail of the calculations.

Once we realize the gargantuan cost of building such "Ultimate Weapons," we are also struck by the repercussions of repeatedly blowing them up! Is that a financially wise thing to do?

Zachary Feinstein, from Washington University in St. Louis, published a paper on December 1, 2015, entitled "It's a Trap: Emperor Palpatine's Poison Pill." Feinstein explains that he first modeled the state of the economy of the Galactic Empire. He then worked out the economic repercussions from the Battle of Endor, when the second Death Star met its end. His conclusion reads:

> *"[Our] picture of the economic repercussions from the Battle of Endor [suggests that] the Rebel Alliance would need to prepare a bailout of at least 15%, and likely at least 20%, of GGP [Gross Galactic Product] in order to mitigate the systemic risks and the sudden and catastrophic economic collapse. Without such funds at the ready, it's likely the Galactic economy would enter an economic depression of astronomical proportions."*

The key point is that the destruction of the Death Star would have a catastrophic effect on the galactic economy. Thousands of jobs were created and maintained by the Death Stars, and their destruction would create decades of extreme poverty and starvation. The rebels simply did not account for that in their attack, on both occasions.

In *Rogue One*, a wayward band of Rebel fighters comes together to carry out a desperate mission: to steal the architectural and engineering plans for the Death Star before it can be used to enforce the Emperor's rule. When armed with the detailed plans, surely the more prudent option would have then been to marshal control of the Orbital Battle Station and turn it over to socially useful production? Had a peaceful plan been adopted for *Death Star I* instead of destruction, it wouldn't have been necessary to waste so many of the galactic resources on building its replacements.

From Death Star to Life Star!

Several useful, and environmentally safer, alternatives present themselves.

Perhaps the Death Star could have been transformed into a mobile science research facility and observatory—a colossal and mobile take on the Solaris Station, the scientific research laboratory hovering near the oceanic surface of the planet Solaris in the 2002 Steven Soderbergh movie *Solaris*.

Or how about a Death Star holiday destination? Sporting slogans such as "Picnic in Paradise" and "The Canteen at the End of the Universe," the Death Star's "satellite city" possibilities could be engaged to move the Death Star to a star system of your choosing. What better way to repurpose the station's crew quarters, five-star luxury options, turbo-elevators, galactic views, and Imperial theater systems?

Less attractively, the Battle Station could have been transformed into a high-security jail to incarcerate the worst elements of the Imperial regime. Off-planet banking is a less populist possibility, providing a tax h(e)aven for the galaxy's wealthy elite.

But maybe the most culturally enlightened alternative would surely be the Death Star Discovery Center, a highly mobile art museum and library. Also equipped with a seed bank, a repository of rare species to protect galactic biodiversity, the Discovery Center could tour the galaxy, exhibiting cultural artifacts from every inhabited planet in the galactic regions.

THE HUMAN FUTURE IS IN SPACE: IS THIS THE GREATEST OF ALL STAR WARS MESSAGES?

American science fiction author Larry Niven was once quoted as saying: "The dinosaurs became extinct because they didn't have a space program."

Niven was being quoted by Arthur C. Clarke, co-writer of *2001: A Space Odyssey*, in conversation with Buzz Aldrin, the second man to walk on the Moon. The two futurists were talking back in 2001 at Clarke's home on the island nation of Sri Lanka. The subject matter stretched to Mars and beyond, as they contemplated humanity's real space odysseys in the century that lay ahead.

Both futurists saw Larry's point.

If the dinosaurs had had a base on the Moon or Mars, they would not have died out. Or at least they would have had a greater chance of survival. Those raptors were smart. So, given our future on a threatened planet, it might be wise to take our inspiration from Star Wars.

A space program had enabled humans from the Core Worlds to found a Galactic Republic, the ruling government of a galaxy that would remain in power for twenty-five thousand years. The Galactic Republic was an aggregate of peoples on different planets, ruled over by a benevolent government. Spread out. Distributed. Not so prone to a single catastrophe, such as a rogue asteroid, or an attack from a Death Star.

Clear and Present Danger

There are many ways in which we can foresee the end of the world. And by "The End of the World," we are not talking about the most popular cantina on Koda, or the space station in the Outer Rim Territories. However, each of the doomsday scenarios *does* have a parallel with a storyline in Star Wars.

Consider the prospect of Earth being hit by a comet. This major impact event releases the same energy as several million nuclear weapons detonating simultaneously. And that's when a comet of only a few kilometers across collides with a planet. The last known impact of a truly huge object (ten kilometers or six miles) was the Cretaceous–Paleogene extinction event sixty-six million years ago, of course. That ruffled a few feathers, or at least dinosaur skins. Clearly, impact events of that magnitude are relatively rare, but Earth's long history is nonetheless riddled with them. Such a cometary impact would be akin to devastation that would be caused if the Earth was zapped by the Death Star.

Smaller impacts, around the size of one or two kilometers across, strike the Earth every half a million years or so. These are often called "threshold global catastrophes." That's because they could cause food chains to collapse, both on land and at sea, by producing so much dust that photosynthesis would cease. Threshold impacts could also cause mega-tsunamis, or global forest fires.

And it would be no good going for the "Hollywood option" by just blowing up the oncoming comet. The gung-ho nuking solution would simply make smithereens of the space rock, and result in it cascading down to Earth, in the way the Death Star should have done on Endor.

Sunburn

If comets don't get us, the long slow burn of the Sun surely will. Eventually.

Just as we have limited resources here on Earth, the Sun, too, has only a certain amount of fuel to burn. For most of its life, the Sun burns hydrogen.

However, the hydrogen will run out, and the Sun will turn to helium for its fuel. And when stars burn helium, it comes with a health warning: the Sun will grow massively in size to become a red giant.

H. G. Wells envisaged this entropic end to the solar system in his famous 1895 novel *The Time Machine*. When Wells's story reaches the end of time, we find an Earth locked by tidal forces. The planets spiral toward a red giant Sun, which hangs motionless in an endless sunset. The solar system is in meltdown.

If scientists have calculated their sums right, something like this will actually happen. In the distant future, perhaps as much as five billion years, our Sun is expected to become a red giant, expanding to two hundred times its present size. When that happens, the Sun is expected to swallow up the inner planets, including Earth. It seems there's more of a chance of survival on Mars, so a Star Wars outward urge will become a necessity.

The Four Horsemen of the Sixth Extinction

Human activity on Earth is sometimes called the sixth extinction.

Scientists count just five mass extinctions in the huge expanse of the last 450 million years, but warn we may well be entering a sixth. In the past, it has taken life up to thirty million years to recover from an extinction—that's over one hundred times as long as humans have been telling tales by firelight. The impact of this still-avoidable sixth extinction would be more like a sci-fi scenario.

Fundamental changes on a planetary scale have begun. The Earth has already met with global warming, acidified oceans, and mass extinction, well before humans first set foot on the planet. Yet, no one knows how long these more recent changes will last. They could be a brief, unique transition in Earth's history. Or they could evolve into a new, geologically-protracted planetary state.

The myth of the Four Horsemen of the Apocalypse comes from the Book of Revelation in the Christian Bible. The vision is that the four horsemen—beings riding out on red, white, black, and pale horses—will bring apocalypse upon the world as harbingers of the Last Judgement.

Experts now believe there are four factors of our current extinction crisis, which make it unique in Earth's history. These four factors are: the spread of non-native species around the world; a single species (humans) taking over a significant percentage of the world's primary production; human actions increasingly directing evolution; and the rise of the technosphere—that vast, sprawling fusion of humanity and its technology.

The Space Age

The evidence is clear.

To better ensure our survival, we need a space program. Like the humans from the Core Worlds in Star Wars, we need to establish a Galactic Republic. Star Wars still presents the most populist example of a distributed human population in space.

We've made a start. The Space Age is already upon us. The evolution of human society has passed through a Stone Age, Bronze Age, and Iron Age. The industrial revolution of the nineteenth century was often known as the Machine Age. However, the Space Age began in 1957 with the launch of *Sputnik*, Earth's first artificial satellite. The first living creature, the dog Laika, was launched into space, also in 1957, followed by the first man in space, Yuri Gagarin in 1961, and the first woman, Valentina Tereshkova, in 1963. And in 1965, Alexei Leonov became the first human to walk in space.

We've learned how to have a constant presence in orbit around the Earth, and from the experience of being there. The *Mir* space station served as a microgravity research lab in which crews conducted experiments in biology, physics, astronomy, meteorology, and spacecraft systems with a mission aim of developing the technologies needed for the permanent occupation of space. Today, the *International Space Station* is the product of lessons learned after decades of launching and operating such space stations.

The Human Future in Space

The creativity of Star Wars is an inspiration to many.

But humans need to get organized, whether the mission is to colonize the Moon or Mars, start mining in deeper space, or turn an asteroid into a space city. New modes of space living and propulsion need to be considered, researched, and developed.

Around the time *A New Hope* was released, a group of professors met for a period of ten weeks to brainstorm the design of space colonies. Their recommendation? Purpose-made colonies for orbit around planets or moons. The colony would be a wheel-like habitat 1.6 kilometers (one mile) in diameter. Colonists would live in a tubular wheel structure, which would rotate to simulate gravity, and use mirrors for solar power. It's not exactly the Death Star, but it's a start.

Scientists have also looked into the idea of building spacecraft that are propelled by nuclear power. Nuclear fuels are lighter and more efficient, and could take a craft much further into deep space—but they are not always safe to use. A further option would be to use an ion engine, in which electrically charged particles are focused into a beam. The beam creates pulses of thrust, which push the spacecraft in the direction of its destination.

NASA may not yet have a hyperdrive. Contrary to propaganda, the Chinese are not building a Death Star. But we humans are all doing what we've done for over half a century—sending spacecraft out into the solar system and beyond, and gazing into deep space with telescopes that peer at galaxies far, far away.

EINSTEIN'S $E=MC^2$ AND STAR WARS: WHAT ARE THE CHALLENGES OF LIGHTSPEED TRAVEL?

Let's imagine Rey is soaring through the sky of Jakku, at night. No starship, no speeder; just Rey. Somehow. Perhaps she's salvaged some kind of repulsorlift tech, and is wearing it around her waist. She's really tearing through that night sky. In fact, she's traveling at the speed of light. Don't worry too much how she got to that speed. This is Star Wars, after all. She cuts a pretty mean figure, of course, as she rips through the planet's atmosphere, so she decides to check her reflection. She pulls a hand mirror out from her expedition backpack, and stares at the glass, knowing the light from the two moons of Jakku is enough to spy her reflection.

But what does Rey see?

If she's moving at the speed of light, then the moonlight reflected from her face couldn't catch up to the mirror. That's because *Rey* is moving at the speed of light and so is *the mirror*. Rey is essentially sitting on top of the light wave, so the light from her face can't catch up to the mirror!

That's a bit weird, isn't it? Rey's image would disappear, vampire-like. In some cultures, vampires don't have a reflection and don't cast a shadow, maybe as a manifestation of the vampire's lack of a soul. But this isn't Bram Stoker. Let's call this a Star Wars thought experiment.

The Experiment Continues

The light on Jakku should *not* disappear.

Whether the starlight is pure, or reflected as moonlight, Rey's image should strike the mirror. However, let's now think through the consequences of her image *not* disappearing. Imagine Finn is also on Jakku, on the outskirts of the Imperial Research Base in Carbon Ridge.

Finn has picked up some of Rey's scavenging skills and he's managed to salvage some macro-binoculars, the field glasses we see Obi-Wan using on Geonosis. Finn's got the special edition specs, the model specialized for use at night. Those macro-bins are good. They allow the user to see distant objects. Some models even enable you to see into space from the surface of a planet, allegedly.

Finn spies Rey through the macro-bins.

He places the two cushioned eyecups against his face, adjusts the top-mounted rangefinder, and uses the readout that feeds data on distance and elevation. He sees Rey, moving at lightspeed, and he sees her reflection in the mirror. Just to be sure, Finn uses the record facility on the macro-bins, and plays back the moving images.

But this is also strange, isn't it? Finn is seeing the light leaving Rey's face at twice its normal speed!

If Rey is moving at 186,000 miles a second, and the moonlight leaves her face at 186,000 miles a second, then relative to Finn, the light should be moving at 372,000 miles per second!

Something's Gotta Give

The solution to the puzzle is rather simple. And it's this: the speed of light, leaving Rey's face, has to be the same for both observers, Rey and Finn.

The only trouble is, the consequences of this solution are not simple. A similar thought experiment led Einstein to conclude that, if everyone (including Rey and Finn) were to see the same velocity for light, then *all* other conventional ideas associated with light motion must change. Classical ideas about time, length, mass, and speed *all* have to be tossed into the dustbin of history.

So Einstein came to the conclusion that light always moves with a definite speed, also known as "c." And for us that means—whether it's Rey or Finn perceiving it—the speed of light will be the same. That conclusion has a startling repercussion for Star Wars: There's no such thing in nature as an instantaneous interaction. In other words, every interaction takes time to get from one place to another.

That's where the deeper difficulty with Star Wars steps in. What would happen when a spaceship approached the speed of light?

Energy and Inertia, $E = mc^2$

When a spaceship picks up speed, it accelerates.

This acceleration is proportionate to the force applied, which means the bigger the force, the faster the ship picks up speed. However, the greater the mass of the spaceship, the harder it is to get it moving faster. This is due to a quality of mass known as inertia. Inertia is the property of mass by which it continues in its existing state. Its resistance to motion, if you like.

Inertia is why it's easier to get the Naboo Royal Starship rolling than it is a Trade Federation battleship.

Einstein argued that as you give an object, such as a starship, more and more energy, rather than going faster, the starship just gets heavier and heavier. Consider the *Millennium Falcon*: To speed along, it needs energy. And to travel at the speed of light, the amount of energy it needs to propel it might swell to infinity! In short, to move the *Millennium Falcon* at lightspeed might take all of the energy in the universe. Just a small snag.

This is clearly the reason the hyperdrive was a necessary invention.

IS THE DEATH STAR A GOOD DESIGN FOR A SPACE STATION?

It's a city-sized behemoth of a battle station, built with the sole intention of breeding terror in the enemy and devastating whole worlds.

Every has to start somewhere, though—from the early ponderings of some progressive thinker, all the way through to the technical accomplishments needed to achieve it.

So where do we stand in terms of space station design?

The closest we've come to even remotely approximating a Death Star would be the numerous space stations that have been put into orbit.

But how do our space stations compare to the Death Star?

Battle Stations

Back in 1977, when Star Wars was released, the US had a functioning space station called *Skylab*. However, by the release of *The Return of the Jedi*, featuring the bigger and badder *Death Star II*, the NASA-led *Skylab* had been grounded.

Unlike *Skylab*, the Death Stars were for military purposes. However, *Skylab* wasn't the only—or the first—space station in existence. By the time of the film's initial release, the Soviet Union had already launched many of their Salyut space stations. Just like the Death Stars, most were actually part of a military space station program called Almaz.

They even had a weapon. It was a Rikhter R-23M "space cannon" that could fire twenty-three millimeter rounds at a potential rate of thirty rounds per second. It was fixed in position, which meant the whole space station had to be reoriented to fire at a target. Ring any bells?

They tested the space cannon before the station's reentry in 1975. They found that, in addition to vibrating the whole space station, the recoil from its firing had to be balanced by employing the station's thrusters in the opposite direction. Stepping up from the Rikhter design, there have been many proposed designs for larger space-based weaponry. Particularly, these ideas were developed for the purpose of defending against potential nuclear missile attacks. Proposals included lasers and ballistic weapons, as well as mirrors that could reflect ground-based laser beams back to another point on Earth.

Since those early attempts to weaponize space, and the fizzling out of the Cold War, militarization has been less of a concern for space station endeavors. Cooperation became the preferred mode of conduct, signified by the development of the *International Space Station* (ISS).

Power Stations

To state the obvious: a space station needs power to run.

The Death Star is said to have used hypermatter reactors to power its systems. Unfortunately, we don't have such exotic technology.

Instead, the ISS uses eight solar array panels to produce 84–120 kilowatts of power. Each solar array panel is thirty-five meters long and twelve meters wide. In addition to solar power, the ISS can also use chemical energy from its rechargeable batteries, which have a life span of about six-and-a-half years.

Beyond that, the most advanced method we have of energy production uses nuclear fission. Nuclear fission is a million times more bountiful than extracting energy from chemicals. Yet, from a given mass, nuclear fusion could produce three or four times more energy than fission. However, scientists and engineers are still trying to figure out how to use nuclear fusion for effective energy production.

In the distant future, possibilities for power production could include antimatter. A kilogram of antimatter reacting with matter can liberate a similar amount of energy to the Tsar bomb, the biggest thermonuclear bomb ever made. However, a gram of antimatter would take one hundred billion years to make with current technology.

Thermal Control

Space station temperatures need to be regulated. On the ISS, the Sun-facing side could get as hot as 121°C, while the dark side can plummet to -157°C. Insulation can limit these extremes by slowing down the time it takes for the ISS to absorb or lose heat. On the whole, cooling down is more of a problem than keeping warm, though.

Space stations generate heat internally due to life-forms, power systems, and technology use in general. The Death Star famously had thermal exhaust ports to vent some of the heat from its reactor. In this case, however, the ports would have to actually be venting some form of matter in order for the heat to escape.

This is because in the vacuum of space, heat cannot be carried away by convection or conduction, both of which require a substance to act on. Instead, the energy has to be radiated away as electromagnetic waves.

On the ISS, an Active Thermal Control System is used where water passes through cold plates that absorb the excess heat. This water would freeze if exposed directly to the cold of space, so another, ammonia-filled loop is used, which carries the heat to radiators that are exposed to space. This allows the heat energy to escape as infrared radiation.

Now, if you want to lose heat, a sphere isn't a good shape. This is because it has the smallest surface area compared to volume. To help cool the massive Death Star, increasing its surface area could be a good start. A way to do this would be to have a convoluted exterior, similar to the surface of a human brain.

Up close, the Death Star looks like a city covered in crevices, towers, and miscellaneous structures. It's possible that some of those features are used for the purpose of radiating excess heat from the Death Star. However, for the sake of efficiency, the heat could equally have been repurposed as an alternative source of power.

Orbital Characteristics

The ISS whizzes around the Earth once every ninety-three minutes. It's the size of a football field, but when we look up we see it as a tiny star-like light moving across the sky. It orbits between 330 to 410 kilometers above Earth's surface in what's called low Earth orbit (LEO).

Objects in LEO experience drag from the outer atmosphere, which slows them down and reduces their altitude. As such, the ISS must conduct frequent boosts to realign its orbit. This can be done using the chemical engines of the space station's Zvezda module, as well as through docked spacecraft. It is hoped that ion and plasma thrusters will help to boost the ISS in the future as they are more fuel efficient than chemical thrusters, which would help to reduce costs.

Like the ISS, the Death Star can also adjust its position using ion thrusters spanning its equator. It also has hyperdrive generators for getting to other star systems. The ISS is strictly intended as an earthbound station, so it only needs enough propulsion to perform its orbital adjustments.

When the Death Star is in orbit above Endor, it stays in the same spot above the surface. On Earth this is known as a geostationary orbit, popularized by previously mentioned science fiction author Arthur C. Clarke. Communications and weather satellites tend to utilize these orbits.

On Earth, geostationary orbits tend to be at a distance of just under 36,000 kilometers. If the *Death Star II* were in geostationary orbit around the Earth, we would see it as an object half the size of the moon in the sky. However, Endor is a moon with a smaller mass than Earth. Therefore a geostationary, or more correctly an "Endorstationary," orbit would actually be closer to the planet.

Replicating Gravity

To maintain its crew's well-being, a station needs life support systems for oxygen, pressure, temperature, and water. However, reduced gravity can also lead to various health problems such as bone and muscle atrophy. To try and mitigate these effects, astronauts on the ISS have to follow an exercise regime.

Earth's gravitational attraction is due to its huge mass, but the space station only weighs about 420 tons, which is too small to have noticeable gravity. Depending on its actual mass, this may not be the case with the Death Star.

Assuming *Death Star I* has a diameter of 75 miles and a mass of 134 quadrillion tons (an approximate measurement previously used in the chapter "How Much Would It Cost to Build a Death Star?") we can work out the gravitational force at its surface. It turns out that a person weighing approximately seventy kilograms would feel a gravitational pull of about a quarter that of Earth's. This would be more than the gravity felt on the moon (which has one-sixth of Earth's gravity).

However, this is only near the surface and would reduce as you descended into the Death Star.

Another way to produce a gravitational-type force is to spin the craft like a merry-go-round. As it spins faster, a merry-go-round becomes harder to stay on, causing riders to get flung towards the outer edge. If a wall were placed at that outer edge, the rider would feel pinned to that wall. These principles can be applied to a spacecraft.

On a spacecraft, that pinning against the wall would feel similar to the way gravity acts to pin us down. The wall would effectively become our down, and the center of the spin would become our up.

Using this idea, engineers have proposed the creation of a giant bicycle wheel habitat. First publicized by Wernher von Braun in 1954, this idea was also featured in *2001: A Space Odyssey.*

People would inhabit the rim of the wheel. The center of the wheel would be what they experience as up. Conversely, the gravity-like notion of downward pressure and weight would be felt in the outwards direction. The same principle would work if a large cylinder shaped station were spun like a rolling pin.

On a sphere like the Death Star there would be difficulties, though. Anything at the equator would move faster, with a greater outward force, than something near the poles, which would be spinning through a smaller distance in the same time. Therefore, the gravity-like force would move from maximum near the equator to almost nothing by the poles.

Gravity on the Death Star comes from another process, though. It uses gravity field stabilizers and compensators that can be turned on or off locally. However, nothing like this can be achieved using current technology.

WHAT MAKES THE MILLENNIUM FALCON SUCH A PRIZED SHIP?

Although it's not the slickest looking vehicle in the parking lot, its appearance is unmistakable. It has a side-mounted cockpit and a rather non-aerodynamic radar dish on the roof. This machine doesn't look like it's built for speed but whoa does it go.

Lando calls it "The fastest hunk of junk in the galaxy," while Leia describes it as a "bucket of bolts." By the way people refer to the craft it wouldn't sound like the best candidate for a spaceship, especially alongside such more graceful ships as the Naboo Royal starship or the H–type Nubian yacht.

So why is Han Solo so fond of this ship?

A Smuggler's Ship

A ship's value depends on what the ship is to be used for, which for Han Solo was smuggling.

Real-life sea smugglers of the past were usually seafaring men with a penchant for adventure or a keenness for a quick profit. Considering Han won the *Millennium Falcon* in a bet, he fits the bill perfectly.

As a smuggler, his needs are particularly different from those of a normal legal trader. Normal traders may go for big containers and ships, but smugglers benefit from contraband in much smaller sizes, which is much easier to load, unload, and conceal. This goes alongside having a maneuverable ship, capable of getting in and out of ports quickly. Eighteenth-century smuggling ships were custom-made with added adaptations to maximize speed, as well as carriage guns and smaller swivel guns to increase their chances of survival.

The *Millennium Falcon* was a modified Corellian YT-1300 light freighter so it was built for carrying cargo. However, it contained hidden storage spaces for smuggling contraband. Han also made many other "special modifications" to the *Falcon* to aid in his clandestine operations. Just like those early smuggling ships, the *Falcon* also had multiple weapons but in the form of quad laser cannons and a retractable blaster cannon, aka a "ground buzzer." It also had homing concussion missiles that were enveloped in an energy shield, used by Lando to destroy *Death Star II*.

Alongside its standard ray and deflector shielding, it has an extremely tough armored hull, which is further reinforced around precious areas such as engine and crew compartments. In preparation for the worst-case scenario, it also has five escape pods.

So the *Falcon* has a few great traits as a vessel built for cargo and adapted for smuggling, and the fact that the famed ship doesn't look as capable as it actually is can only be of benefit to a smuggler, who ideally would like to travel around freely and unnoticed while using their particular knowledge and ingenuity to overcome challenges.

As such, we can't underestimate the influence of a highly skilled and well-motivated pilot. This was just as important for the eighteenth-century sea smugglers as it is for the interstellar space smuggler. It's also evident in Rey's ability to fly the *Falcon* so well when she steals it to escape with Finn in *The Force Awakens*, even though she described it as "garbage" when she first saw it.

How can you judge between a good ship and a "piece of junk"?

Don't Judge a Book by Its Cover

It makes sense that people in the Star Wars universe judge the *Millennium Falcon* on its looks. It's not just that it looks like it needs constant repairs; it's also because it looks like a regular freighter. It's like putting a cargo boat next to a yacht and asking which is best.

However, despite its appearance, "this baby's got a few surprises left in her." So what are her redeeming features?

Capacity: it's a cargo ship!

On Earth, cargo ships have been around for millennia but in 1978 the first ever space freighter went into use. It was an unmanned resupply spacecraft used to transport vital supplies to space stations in low Earth orbit. This means they only needed to get a few hundred kilometers above Earth and required no life support systems.

In contrast, a freighter like the YT-1300 is manned and has to travel light-years between star systems. Meaning it needs extra features such as a cockpit, crew quarters, and hyperspace engines; all of which affect the design and capacity of the ship. It also had padded walls and a recreational area with a hologame table and sofa.

Appearance: it's a flying hamburger!

Some craft need to have a certain shape and color to give them stealth-like qualities such as the B-2 bomber. Nothing about the *Falcon* cries "stealth," though, especially when it's considered too small to have a cloaking device. The closest it came to being stealthy was when Han Solo flew straight at the Imperial Star Destroyer *Avenger*, then clamped on to the hull for a little while before detaching and floating away when the *Avenger* released its garbage.

The *Millennium Falcon* frequently flies within atmosphere where aerodynamic forces come into play, particularly drag. A major issue is the protruding cockpit and sensor dish that would have to be structurally strong to withstand the drag forces exerted on them when moving through air at speed.

The effects could be mitigated if the ship's shield also created a region around the ship that acted as an aerodynamic surface instead of the ship. However, the shield generators would take the brunt of all the aerodynamic forces on the shield, so it would need to be extremely robust, too.

Propulsion: it has two sets of engines for main propulsion as well as repulsorlift engines and landing jets.

Its Girodyne-manufactured sub-light engines propel the *Falcon* at speeds slower than the speed of light. It's said to work via a fusion reaction that breaks down fuel into charged particles, which are ejected from the ship. To provide the necessary thrust for the *Falcon*'s maneuvers, this would require either a great deal of propellant, or else it would have to be ejected at ridiculously high speeds.

Thrust vector plates above and below the exhaust allow the thrust to be directed upwards or downwards to change the orientation of the ship. This is similar to the gimbaled thrust system used in modern rockets and jets.

Using the sub-light engines the *Falcon* can apparently achieve a speed of 1,050 kilometers per hour through atmosphere. For comparison, that's less than half the speed of an F-14 Tomcat, so it's not pushing the envelope there.

For longer journeys, made through hyperspace, the *Millennium Falcon* uses an Isu-Sim SSPO5 hyperdrive engine. This unit has had numerous tweaks and upgrades made to it to bring it up to a class 0.5 hyperdrive. The lower the number, the faster the hyperdrive. Star destroyers tend to use class 2.0 hyperdrives, which is why they are no match for the *Falcon* through hyperspace. The *Falcon* has even made the jump to hyperspace from inside the hangar of a ship.

Navigation: it made the Kessel Run in less than twelve parsecs!

In case you didn't know, distance can have many units; one of which is the parsec.

A parsec is a very long distance equivalent to how far light can travel in 3.26 years, which is nineteen trillion miles. That's more than five thousand times the distance to Pluto.

So Han Solo's claim is that the *Millennium Falcon* made the Kessel Run in less than 230 trillion miles. As an indication, this is 4.5 times further than the brightest star in the sky, Sirius. This is a distance that can only be achieved through hyperspace.

A spaceship needs a good computer to handle its many complex functions, but when it comes to navigating your way through hyperspace, computing power is vital. This is to avoid regions of space that would be hazardous, such as planets, stars, or black holes. It's like having a good satnav.

Your satnav can advise you of the quickest route to take, knows what time you'll probably arrive, warns you about traffic delays and soon might be offering you the scenic route. It basically has a database of options for your journey; you just have to tap in your destination.

The *Millennium Falcon's* navigational computer, or navicom, is basically an interstellar satnav. Instead of using satellites to ascertain position, it may use a method based on the positions of detectable stars, along with the ship's speed relative to some defined celestial object. It would also require up-to-date maps and data on the galaxy, alongside a good bit of software to crunch all the data.

We already know that the *Millennium Falcon* has undergone special modifications, so the computer's probably sorted. Regarding maps, it's possible that Han Solo and Chewie may have acquired a database of special routes through hyperspace that other smugglers or mainstream databases do not have. A bit like having a satnav designed by taxi drivers, covering the best shortcuts in the neighborhood.

When there's more than one way to travel from A to B, calculating the shortest route will be a big factor on who can get to B the quickest. It's possible that this is how the *Millennium Falcon* made the infamous smuggler's Kessel Run in less than twelve parsecs.

All in all, the *Millennium Falcon* may look like a hunk of junk, but that's testament to its robustness and the amount of refits it gets. It's these refits and "special modifications" that have given the *Falcon* the edge over much of its competition.

It is a prized ship indeed.

WHAT ARE THE CHANCES OF NAVIGATING SAFELY THROUGH AN ASTEROID FIELD?

"Sir, the possibility of successfully navigating an asteroid field is approximately 3,720 to 1."

This is a classic quote by C-3PO. But Han Solo doesn't want the odds; he just wants to survive, and he's willing to take on the most improbable situations to make that happen.

But how accurate were C-3PO's odds anyway? Especially considering that the crew did in fact manage to successfully navigate an asteroid field. How is it even possible to make a prediction like that?

Taking a Gamble

Han's a gambling man. It's the reason he has the *Millennium Falcon* in the first place. When he opts to go through the asteroid field C-3PO tells him the odds to indicate how absolutely crazy he must be and try to persuade him otherwise. But do the odds really give that impression?

Consider this: in the UK, 70% of adults play the national lottery even though the odds of winning the jackpot are now forty-five million to one. So are people who play the lottery crazier than Han Solo? Not exactly.

Although many people hope to win the jackpot, there's more than one way to achieve success in the lottery. It's not just based on a single outcome of win all or lose all; there are other prizes up for grabs.

The chances of achieving the most minor win are actually a more reasonable 96 to 1, and the prize that follows after that is 2,179 to 1. Therefore, you could argue that if Han is irrational for taking on odds of 3,720 to 1 against, then people who play the lottery are *less* irrational than he is.

There's another factor, too. The cost of the gamble. Losing a game on the lottery would cost you a couple of dollars, whereas Han Solo is gambling with his life . . . as well as Leia and Chewbacca's. If people in the UK were putting their lives on the line every time they played the lottery, you'd most likely find that it is nowhere near 70% of the population who would be willing to take the gamble.

In fact, most people wouldn't play a game of Russian roulette, which gives the comparably better odds of 5 to 1 in favor of succeeding. That's an 83% chance of survival! In comparison, C-3PO's stated chances of success are 1 out of 3,721 or 0.027%. This is about 3,090 times worse off than playing Russian roulette. Putting it this way, Han Solo is one crazy dude!

What Are the Chances?

When we make choices, we normally deal with likelihoods rather than certainties. Probability is a measure of the likelihood that some event will happen. In this case the event is successfully navigating an asteroid field.

When you flip a coin, there are only two possible outcomes. In a fair trial (i.e., one with no hidden biases or trickery), each outcome would have an equal chance of occurring. It's called fifty-fifty, as each outcome should occur 50% of the time.

Imagine you have two horses entering a race. If all other things were equal, they should both have half the chance of winning. But all things aren't equal. All sorts of variables come into play. The horses are different and the jockeys are different, with varying levels of skill, training, and talent; even the weather can have an effect on the outcome.

To calculate reliable odds of some event occurring, one must acquire knowledge of how each of the different factors can affect the chances of a successful outcome. This applies whether it's a race, a game, or even the chance of snow on Christmas. The person who makes this judgment is called the odds compiler. Using their knowledge of a particular sport or event, they are able to work out how likely a particular outcome is.

The best way for them to do this is to look at how an event had played out previously. For C-3PO this would mean retrieving as much information as possible from past events that matched the present one. The result of the search may have indicated that for every ship that has successfully navigated an asteroid field, there have been 3,720 that have not.

However, this calculation doesn't take into account the skills of the individual pilots, the size and maneuverability of the ships used, or the circumstances in which they entered the asteroid field. Even if these factors were known, they would need a probability assigned to each of them, which without sufficient data would be problematic for a machine, like C-3PO, to calculate.

This is why a major skill of an odds compiler lies in their ability to get a feel for a situation through experience and intuition. This is something that C-3PO, as a robot, lacks. It can't comprehend the subtleties of human ambition and the effect it can have on success. In this way, any odds compiled by C-3PO would always be off the mark, neglecting the most vital part of the equation: the human aspect.

So what exactly were they getting themselves into?

"That Wasn't a Laser Blast. Something Hit Us!"

In human space exploration, asteroid-filled regions aren't referred to as asteroid fields. They're generally described by the space they occupy or by a particular grouping, such as the Near Earth Asteroids, Jupiter Trojans, or asteroid belt.

When it comes to successfully navigating through a region of asteroids, their size, velocity, and separation are the most important factors. The asteroids in the Hoth field are extremely close together. This could imply that it is newly formed, or that they're just flying through a region where two huge asteroids have recently collided.

The majority of asteroids in our solar system lie between the orbits of

Mars and Jupiter, within the asteroid belt. The asteroids range in size from just a few meters to more than nine hundred kilometers. Over time they collide, breaking into smaller chunks and getting further apart.

As such, there are hundreds of millions of asteroids, with about twenty-five million larger than one hundred meters and more than two hundred bigger than one hundred kilometers. However, it's estimated that the objects in the asteroid belt would only add up to about 4% the mass of the moon. If you think about distributing that material along a belt that goes right around the Sun, you can see that they must be pretty spaced out.

The average distance between asteroids is said to be about 965,000 kilometers. This means they are two times further away from each other than Earth is from the moon. Therefore, the chances of navigating through our asteroid field are high.

Indeed, since *Pioneer*, which was the first probe sent through the asteroid belt, all subsequent spacecraft have successfully made it. In fact, based on existing data, the chances of success would be 100%. NASA has estimated the likelihood of a space probe hitting an asteroid to be about one in a billion.

However, our asteroid belt isn't the only one out there, so other arrangements probably exist. For example, the Epsilon Eridani system has two asteroid belts, with the outer one denser than the inner.

So where does this leave Han's chances of successfully navigating an asteroid field?

Navigating Probabilities

When C-3PO blurted out the odds, it clearly didn't comprehend how good or lucky a pilot Han was. Additionally, the psychological variables that could affect Han's ability to focus and perform well are far too unpredictable.

More predictable for a robot like C-3PO would be the plotting of asteroid trajectories. Technically, it should be better at Han at navigating an asteroid field, if its sensors were sufficient. An even better solution would have been to provide the ship itself with an early warning system, which could have stopped them from getting into the mess in the first place.

A similar system was set up for the detection of icebergs on Earth. After the sinking of the *Titanic*, the International Ice Patrol was set up to monitor the danger from icebergs in the North Atlantic Ocean. They provide the maritime community with relevant iceberg warning products.

We also have asteroid detection on Earth. NASA's Planetary Defense Coordination Office is responsible for ensuring the early detection of potentially hazardous objects (PHOs)—which are asteroids and comets that may come within a certain distance of Earth and be big enough to reach the surface. They track and characterize their size and trajectories to make predictions on their orbits.

However, if C-3PO were responding to the present situation, it would have probably calculated the probability of navigating *this* asteroid field rather than *any* asteroid field. It was likely just repeating some statistic yanked from its memory banks. However, there was other information at C-3PO's disposal.

For example, the *Falcon* had an armored hull and deflector shields along with good maneuverability, all of which should alter the probability of success. Also, an awareness of the type of asteroids in the particular region would have helped.

It's now believed that many asteroids may not be made from solid rock, but several pieces that have coalesced. They're called rubble piles. If a shielded *Millennium Falcon* were to have a glancing blow, it might just gouge out a bit of the rubble or just crash into the asteroid without suffering major damage.

So the real odds of navigating a randomly encountered "asteroid field" are probably impossible to calculate without adequate knowledge of all the factors that could hinder or encourage success. One thing's for sure, though, if it's anything like our own asteroid belt, the chances of success are pretty high.

HOW LIKELY ARE THE INTERSTELLAR COMMUTES IN STAR WARS?

Luke Skywalker is in the Dagobah System with Yoda. While he's there, he has a vision that his friends are in trouble, so he hops into his X-wing starfighter and hightails it over to the gas planet Bespin.

A quick look at a map of the Star Wars galaxy reveals that Bespin is at least ten thousand light-years away from Dagobah. This obscene distance doesn't deter Luke in his cramped cockpit. He manages the journey in what seems like a matter of hours, if even that.

Now, of course, Einstein has informed us that objects can't travel faster than the speed of light. Clearly, this isn't an obstacle for the inhabitants of the Star Wars galaxy. They don't follow the typical rules of science. They just take a shortcut through hyperspace.

This option isn't available to all ships, though, and sadly for the rest it's just plain old sub-lightspeed travel. Unfortunately for us, we fall into this group, too.

Given what we currently know about space travel, how likely are Star Wars-style interstellar commutes?

Baby Steps

Travel to other planets can take many months or years, but to achieve interstellar travel to even the nearest star systems would take many decades or centuries.

The furthest humans have traveled from Earth was on the 1971 Apollo 13 mission when three astronauts swung around the far side of the moon. They were 400,171 kilometers away from Earth and had done so in a flimsy Lunar Module called *Aquarius*. Their round trip lasted just under six days, using liquid fuels for chemical propulsion.

The mission wasn't a bed of roses, though. Many things went wrong, and it's a testament to human ingenuity that they managed to get back safely. *Apollo 13* remains as the space mission that carried humans the furthest from Earth.

It's amazing to think that in the almost fifty years since the Apollo missions, we have still come no closer to stepping foot on any other world, let alone another star system. We are a very, very long way from interstellar travel.

The major problem is how to support life on such extreme journeys, through such a hostile place. Even a six-month journey to Mars poses major difficulties; imagine trying to leave the solar system.

That being said, we have managed to send something that far. It's an unmanned spacecraft on a journey that has now surpassed interplanetary status. The *Voyager 1* space probe, launched in 1977, is now more than twenty billion kilometers away (134 times the distance from the Earth to the Sun), speeding through interstellar space.

Presently traveling at a speed of around sixty-two thousand kilometers per hour, it has taken almost forty years to get to interstellar space using gravitational assists and hydrazine-fueled thrusters. In perspective, if it were heading in the direction of the nearest star, Proxima Centauri, it would still take more than 73,700 years to get there.

Recently, there has also been the Breakthrough Starshot initiative, which is supported by Stephen Hawking. It hopes to use a ground-based beam of light to propel light sails attached to ultra-light nanocrafts. The crafts could theoretically reach speeds of one hundred million miles per hour and reach a nearby star system such as Alpha Centauri in twenty years. It could then beam back images and other information.

Clearly, these aren't good enough options for manned interstellar travel, but we've only really been in space for little over half a century.

It's not cheap to get a space mission off the ground, either. Many resources are needed, alongside reliable cooperation and infrastructure.

To give you an idea, the Apollo program required four hundred thousand employees as well as the support of twenty thousand industrial firms and universities to send humans to the moon in 1969, at a cost of twenty-four billion dollars.

Where are we now regarding spaceship technology?

Spaceships

We've had a fairly consistent battle with flight technology. It seems to make its biggest advances through war.

The development of flight changed world transport, with the First World War as a huge catalyst. Subsequent developments in rocket technology also gained prominence through wars. For example, World War II spawned the V-2 rocket, and the following Cold War instigated a space race which led to the creation of the first intercontinental ballistic missiles (ICBMs).

An ICBM would later power human exploration of space by launching Yuri Gagarin into orbit in 1961. Evidently, war has a hastening effect on the development of new technologies.

Less than a decade after entering space, humans set foot on another world, after being launched to space atop a *Saturn V* rocket booster. For half a century the *Saturn V* existed as the biggest rocket ever built, but now NASA's new Space Launch System is set to trump it.

NASA describes it as an advanced launch vehicle for a new era of exploration beyond Earth's orbit and into deep space. However, some planned space vehicles don't require booster rockets to get them into space.

Space planes are intended to take off normally, and enter into orbit or space. These are the types of ships that are most common in Star Wars. There are a few on the horizon, such as the *Skylon*, which is intended for activity in low Earth orbit. Virgin Galactic also hopes to get their space fleet up and flying soon, too.

Propulsion Technologies

Regarding interstellar travel through normal space, there are two Star Wars technologies that have actually been used for space missions.

Ion thrusters, which are used by the TIE fighters, have already been used in communications satellites and space probes such as *Dawn.*

Unlike the TIE fighters, these thrusters perform in a slow and steady manner. They accelerate ions in one direction to produce a tiny thrust in the opposite direction. Their beauty is that they are efficient at turning fuel into thrust, operating steadily and continuously to build up to incredible speeds.

However, their speed is limited by the amount of fuel they have. When one runs out, it cannot accelerate anymore. As such, you wouldn't be able to get to another star system in your lifetime.

In *Attack of the Clones*, Count Dooku makes a journey from Geonosis to Coruscant after battling Anakin, Obi-Wan, and Yoda. His craft, the *Punworcca 116*-class interstellar sloop, uses a solar sail. However, the journey's pretty much halfway across the Star Wars galaxy, requiring the use of the Corellian Run hyperspace route. Therefore, he would still need a hyperdrive to do this journey.

So what good is a solar sail?

In regard to interstellar travel, a solar sail can take you away from a star. It can also benefit from gravitational slingshots around planets on the way out of a solar system. However, as the solar power decreases with distance, it would not achieve speeds great enough to allow quick travel to another star.

NASA is building a solar sail for their Near-Earth Asteroid Scout (NEA Scout) mission due to launch in 2018. Current estimates suggest a speed of 150,000 miles per hour (0.022% the speed of light) could be achieved after three years of operation.

In these crafts, travelers would need to wait for a long time to get up to speed. When they finally did, they would need a fair bit of time, and importantly distance, to slow back down again to land or dock at their destination. Technologies like this in the Star Wars universe just wouldn't cut it.

Unfortunately, current technologies are more prohibitive of interstellar travel than the technology within the Star Wars galaxy.

Star Wars achieves interstellar travel by using hyperdrive engines. We have no idea whether in the future a way may be found to achieve—or even travel faster—than the speed of light. Apart from some remote possibility of wormholes, the laws of physics currently say it's not possible.

It might just be that most of space is out of direct reach to us, and we are destined to only occupy the nearest stars. However, researchers at Oxford University's Future of Humanity Institute have considered a more universe-spanning future for humans:

> *"Traveling between galaxies—indeed even launching a colonization project for the entire reachable universe—is a relatively simple task for a star-spanning civilization, requiring modest amounts of energy and resources. . . . Humanity itself could likely accomplish such a colonization project in the foreseeable future, should we want to."*

WHY DO STAR WARS SHIPS MAKE BANKED TURNS IN THE VACUUM OF SPACE DURING BATTLES?

We've seen the enthralling space battles that help to bolster the status of Star Wars as a space opera:

Enemy TIE fighters screech in behind rebel spaceships, which bob and weave from laser-cannon fire as if in a dogfight.

At the final push, Luke Skywalker is pursued by Darth Vader and two other TIE fighters, but is helped by Han Solo, who sweeps in to pick off one of the fighters. He banks off to the right, giving Luke the all clear to deliver the critical blow to the Death Star.

The fact that the *Millennium Falcon* has to bank is unusual, as banking is a consequence of flying within atmospheres. So why does any spaceship in Star Wars make banked turns in space?

A Matter of Aerodynamics

Aircraft banking has everything to do with aerodynamics.

Aerodynamics is basically how air moves around things. Earth has an abundance of air, but it tapers off into a relative vacuum in space. This is why aerodynamic considerations only apply to crafts operating on planets with atmospheres.

Wings can create lift because air moves over them, or more specifically, because the wings move through the air. The force that moves the wings and vehicle through the air is called thrust, and there are various ways to produce it.

As a craft moves through the air, it can experience resistance known as drag. Drag is generally the enemy of thrust, but having an aerodynamic body shape can reduce it.

The center of gravity (COG) is the point around which all of a craft's mass is balanced. If you push the craft in a direction that does not go through its COG, the craft can turn, altering its attitude.

The attitude is the orientation of the aircraft relative to some fixed frame of reference, such as a horizon or the surface of a landing bay.

Attitude Adjustments

Attitude is usually defined by three characteristics: pitch, yaw, and roll. Each one is affected by a different control surface, namely the elevators, rudder, and ailerons, respectively. Each is involved with rotation of the aircraft on one of three axes.

Pitch is on the "latitudinal" axis, and indicates when the nose points up or down. Yaw occurs on the "normal" axis and is equivalent to the motion our head makes when we look left and right. Then roll is along the "longitudinal" axis, having the effect of turning the plane on its sides or rolling.

For an aircraft moving in a particular direction, it takes a certain amount of time to turn. This is known as its rate of turn, or radius of turn, depending on how you measure it.

Generally, the faster something is traveling, the more space it needs to turn. So an F-16 Fighting Falcon would have a bigger rate of turn than a slower Cessna plane.

So where do banked turns come into all this?

Banking in Atmosphere

When a craft wants to turn, it doesn't just yaw. That would result in a long rate of turn. Instead, the plane makes use of the larger lift force that is generated by the wings.

In straight and level flight, the total lift force from the wings points straight up, balancing the weight of the craft. When a plane banks or rolls to the side, though, a component of that lift force is now pointing in the direction of the turn. It's this sideways component of lift that pulls the plane in the direction of the turn.

The plane stays in the air because there is a substantial component of the lift that still points up. However, to come out of the turn requires the control surfaces to be employed again.

Now, if you look at the vast majority of spaceships in Star Wars, they don't have the control surfaces needed to change the orientation of their ships. This would make sense, since these ships are built mainly for space operation where the laws of aerodynamics don't apply.

Nevertheless, the ships still manage to turn and even visibly appear to bank like a plane needs to on earth. Clearly, the spaceships must use another means of changing their orientation in space—one that works in that environment despite the inadequacy of wings and control surfaces.

If Wings Don't Work, What Are the X-Wings All About?

The X-wing is a VTOL (Vertical Take-off and Landing) aircraft that can take off vertically and hover independently. They never hover at a great distance above the ground, though, which is possible in most real-life VTOL aircraft. As such, they might use repulsorlift technology to achieve lift.

On the X-wing starfighters, the wings (or "S-foils" as they call them) were locked flush against each other for normal flight. In this closed mode, they are almost like a classic two-winged aircraft. But in combat, the wings would open up to give the famous X-wing appearance.

On a planet like Starkiller Base, which has an atmosphere, four wings would create additional drag on the X-wing. This would reduce its maximum speed while also changing its maneuverability. So why do it?

Changing wing configuration can allow a craft to alter its flight characteristics. Many real planes have used this capability such as the Concorde, Tornado F3, Grumman F-14 Tomcat, and other swing-wing aircraft.

Using this trick, they can fly faster, obtain better maneuverability, or change to a configuration that achieves longer flight time out of the same amount of fuel, i.e., increase their range. However, these changes are only effective within an environment that has an atmosphere.

The question becomes this: Does opening the wings provide any benefit to space flight? Nope. It turns out that opening the wings just gives the wingtip-mounted laser cannons an increased firing range, which makes sense for combat, but not much else.

The X-wing has half the mass of an F-14 Tomcat and its maximum speed in atmosphere is 652 miles per hour, which is less than half the speed of an F-14. So we'd expect it to have a better rate of turning. However, the wings are too small to produce a good turn and it lacks control surfaces. So the craft must use another way to turn.

How Do Ships Change Direction in Space?

In space, objects just keep traveling in whichever direction they are pushed. For a spacecraft, this push comes from its thrusters, which propel material in one direction to generate thrust in the opposite direction. This is known as a reaction engine, which obeys Newton's third law of motion.

Another way is to employ a reaction wheel, which uses gyroscopic motion to induce a turning force within the craft. With reaction wheels, the faster or more massive the gyroscope is, the bigger the turning force. Gyroscopes have been used to control attitude in satellites, telescopes, and the International Space station.

When Star Wars ships like the *Millennium Falcon*, X-wings, and TIE fighters want to turn, they don't seem to have separate thrusters. An internal reaction wheel could be a candidate, but it would have to be incredibly massive or strong enough to spin incredibly fast without distorting and flying apart.

A more reasonable solution is thrust vectoring. Thrust vectoring allows the average flow from the engine jet to be deflected away from the center-line of the aircraft, providing a turning force. This is used to great effect on rockets and jet fighters and is also a feature of the *Millennium Falcon*.

Dogfights in Space

The *Falcon* performs rapid, banked turns like a jet fighter. It can even handle almost right-angled turns, as well as flip on its latitudinal (pitch) axis, as demonstrated by Rey while being chased by the First Order's new TIE fighters on Jakku.

Its thrusters are lined along its rear end with thrust vector plates positioned above and below. As these can only move up and down, the ship really only has two degrees of freedom. This is either pitch or roll. However, in an atmosphere, it can also use repulsorlift technology to oppose the gravitational pull of the planet and maybe even to assist in maneuvers.

If the *Falcon* wants to turn, it would first need to roll by having the left and right most vector plates deflected in different directions. Then once it's at the required degree of roll it would pitch, which would point the *Falcon* into the direction of the desired turn. Such banking would be vital to the ability of the *Falcon* to use its thrust vectoring effectively in turns.

For an idea of how effective thrust vectoring can be, check out the F-22 Raptor. The pilots can pull off super-tight turns and extraordinary maneuvers. A computer is needed to handle the precise control of the thrust vectoring, which would likely be the case on a Star Wars ship.

To conclude, even though aerodynamics aren't a factor in space, ships like the X-wing and *Falcon* could achieve sharper turns by banking when using thrust vector control.

PART II
SPACE

WOULD THE DEATH OF THE DEATH STAR SPELL THE END OF ENDOR?

Picture the scene:

An Imperial shuttle with Luke alone in the cockpit rockets out of the main docking bay of the Death Star. The entire section of bay is blown away. Snatching victory from the jaws of defeat, Lando expertly pilots the *Falcon* out of the exploding superstructure and speeds toward Endor, just seconds before we see the Death Star supernova into oblivion.

Lando and Nien Nunb laugh in light relief. We cut to the Endor forest where Han, Leia, Chewie, and the Ewoks gaze up at the sky to witness the Death Star's final flash of self-destruction—little more than a pretty firework powder-puff in the heavens. Everyone cheers. Soon enough, end credits roll over a star field.

Problems? Possibly.

There's little doubt this scene from *Return of the Jedi* is beautifully realized and carefully constructed. Credit where credit's due. But let's take a step back and consider the science of the sequence. What would really have happened to the Ewoks on that forest moon when the second Death Star was so ruinously defeated? Would the satellite world of Endor prove to be the sanctuary its other name implies? Can science help us re-imagine an alternative ending to the movie?

The Death Star as a Moon of Endor

Let's rewrite the scene.

First up, we'll need to consider some situational science and get answers to some technical questions. What are the relative dimensions of Endor and the Death Star? How far apart do they sit in space? How much energy will the Death Star radiate on detonation, and how will it really explode? And how will that explosion make itself felt on the surface of the Sanctuary Moon? All pretty simple.

Yet, from the get-go, there is some uncertainty. How big exactly is *Death Star II*? Canon quotes it as 160 kilometers, which we assume to be its diameter, as the first Death Star had a diameter of 120 kilometers. Other sources suggest that *Death Star II* was about 3% the size of Endor, which has a diameter of 4,900 kilometers. That would make the second Death Star's diameter a mere 147 kilometers.

Other sources say *Death Star II* had a whopping diameter of 343 kilometers.

This last figure is estimated by studying Star Wars at source. On analyzing the hologram from the pre-attack briefing scene from *Return of the Jedi*, the relative sizes of Endor and the Death Star can be reckoned. As we said, Endor is a full 4,900 kilometers across, and the "pixelated" form of the Death Star is about 7% of the diameter of the Sanctuary Moon, giving the figure of 343 kilometers. Far from perfect, but let's go with this estimate, as it does have the benefit of being a primary source of data.

Death Star Sitting Pretty

Now we've worked out the relative sizes of the two bodies, how far apart do they sit?

Knowing the altitude of the Death Star tells us how far Endor is from the blast zone, so we can soon work out how much damage it might take. Gazing again at that hologram from *Return of the Jedi*, it looks like the very centers of Endor and the Death Star are about 2,910 kilometers apart. Subtracting the radius (4,900 ÷ 2 = 2,450) of Endor from this separation

(2,910 – 2,450 = 460) tells us that the Death Star sits at a height of 460 kilometers above the surface of the Sanctuary Moon.

The mass of the Earth can be used to help work out the mass of Endor. The ratio of the two masses, Earth and Endor, is proportional to the square of their radii. So popping in a value for the radius of each body (2,450 kilometers for Endor and 6,371 kilometers for Earth), as well as a value for the mass of the Earth, we find that Endor is roughly 15% as massive as the Earth. That's more massive than Mars. It is true that canon suggests only 8% of Endor's surface is ocean, and rock is denser than water, but Endor's density sits somewhere between that of iron and uranium. Endor is a pretty densely packed little planetary moon! It can take a punch or two.

So let's turn our attention to the Death Star, sitting pretty up there, in the Endorian sky. How's it doing that exactly?

Remember that an energy shield—generated from the surface of Endor—protected *Death Star II* from attack. The shield had to be deactivated before any assault on the space station could be considered. To stay in touch with the shield generator, the Death Star needed to sit in the same spot above Endor. This much is clear in the pre-attack briefing scene. Such an orbit is known as a synchronous orbit, and it's similar to that of communication satellites around the Earth.

It makes most sense to assume that the Death Star is able to hang in synchronous orbit through the use of the same repulsorlift antigravity technology used on speeders, pod racers, and Senate pods. The Death Star repulsorlifts would be massive, of course, as this is some job they would be doing.

The Death Star Explodes

What kind of smithereens would *Death Star II* be smashed into on explosion? And what would happen to those smithereens?

In the recounting of the relevant scene from *Return of the Jedi* we said that the *Death Star II* explosion was little more than a pretty firework powder-puff in the sky. Also, the blast *is* a little lame when compared to the detonation of the first Death Star. *Death Star I* went nuclear immediately after being hit by two tiny proton torpedoes. But the explosion of the second Death Star obligingly allowed rebel pilots to escape its interior before letting loose.

All of this is bad news for the Ewoks.

Thinking about the way in which *Death Star II* explodes, we can assume there won't be much in the line of vaporized smithereens. Just huge scraps of Death Star in supernova. We can also assume that there won't be too much additional velocity on the fallout from the blast either. Once the detonation happens, the smithereens will be traveling much too slowly to stay in orbit.

Thus, it turns out that the powder puff is the prelude to a rain of destruction from above. The entire mass of the Death Star will just fall onto Endor. The focus of the fallout is the site of the shield generator, the location of our heroes and a group of Ewoks.

But how much mass of burning Death Star will there be?

We touched upon this briefly in our calculation of what it would cost to build *Death Star I*. Assuming a similar shell-build for *Death Star II*, and noting its expanded radius of 343 kilometers, a good approximation is a mass of 10^{19} kilograms. This is an interesting number. It's the same order of magnitude as the mass of Saturn's rings—those countless smithereens, tiny and titanic, that make up the bling of the gas giant's signature halo.

We can also work out a speed for the particles. If it were assumed the Death Star made an orbit of Endor each day, then the smithereens would have an initial speed of 212 meters per second after detonation. This is way smaller than the 4,500 meters per second needed to stay in orbit. This speed of 212 meters per second can be used to work out the impact energy with which the Death Star's smithereens hit Endor.

This is one huge impact.

The careening mass of smithereens would hit the Sanctuary Moon at a speed more than 2,800 kilometers per second. That's over 6,000 miles per hour. The impact would create a crater 700 kilometers (435 miles) wide, a full four times bigger than the one left behind by the asteroid that did for the dinosaurs. Endor is only 15% the mass of the Earth.

You can only imagine the extent of the carnage.

Everything on the surface of Endor would be annihilated. The moon's atmosphere would also suffer. It would be broiled by the exploded particles, tearing a path from the blast to the crater. The seas of Endor would flash into steam, as the forests began their long burn into a global firestorm lasting into the night.

The Death of the Death Star, Take Two

So, re-picture the end sequences of the Battle of Endor:

We cut to the deep forest where Han, Leia, Chewie, and the Ewoks gaze up at the supernova detonation of the Death Star. An expanding envelope of fire impinges on the Endorian atmosphere. It's as though every star in the broad expanse has been hurled from its seat, and sent lawless through the wilds of the sky. As thick as scorching snowflakes, thousands of burning meteors are shooting down in every direction, with long trains of light following their course. Every heart is filled with horror at this majestic display of destruction.

Our heroes make their move. Mounting 74-Z speeder bikes, they tear through the forest as the planet burns. Keeping just ahead of the line of fiery destruction, they feel the heat of the forest fires mounting as they make their way to the sanctuary of the *Millennium Falcon*, where Lando waits. Mission abort. The dark side has won. For now. End credits over a star field.

WHAT DOES THE STAR WARS GALAXY TELL US ABOUT OUR OWN MILKY WAY?

Star Wars is replete with aliens.

They include Arkanians, Blood Carvers and Bouncers, Crokes and Ewoks, Gungans and Hutts, Tusken Raiders, Wookiees, and Womp Rats.

All told, there are hundreds of species living on thousands of worlds. However, the geography of the inhabited part of the Star Wars galaxy poses an interesting comparison with our Milky Way Galaxy.

Galaxies

Consider galaxies.

There are more stars in our universe than there are grains of sand on all of the beaches on planet Earth.

It's worth pausing for a while to imagine being on one of those beaches, soft golden sand sweeping off into the distance. Bright sunny day, of course. In fact, let's make it the Caribbean. No expense spared. You reach down, hands cupped, and gather up two handfuls of the golden sand. Then you let the sand fall through your fingers, the grains glistening as they catch the sunlight. Each grain is a star. And each star is a sun, like our own local star, the Sun. You saunter on a few more steps, and again you gather up the sand and let it fall. And so on, over all the beaches on Earth. So much sand, so many stars.

A group of researchers at the University of Hawaii actually tried putting a number to the grains of sand on the world's beaches. The Earth has roughly (and we're speaking *very* roughly here) 7.5×10^{18} grains of sand, or seven quintillion, five hundred quadrillion grains.

And yet, there are ten stars for every one of these sand grains. That's also eleven times the number of cups of water in all the Earth's oceans, and one hundred billion times the number of letters in the fourteen million books in the Library of Congress. (These incredible stats are somewhat put into perspective by the fact that the number of stars in the universe is the same as the number of H_2O molecules in just ten drops of water!)

On a large scale, swarms of such stars dwell in galaxies, effortlessly wheeling their way through the vastness of deep space. Each galaxy contains millions, if not billions, of stars. Under cover of the night sky, some galaxies can be seen with the naked eye. It's worth remembering that the very word "galaxy" derives from the Greek term *galaxias*, or "milky circle," for its appearance to the eye.

Imagine yourself again, on our same trip, under the starry reach of the Caribbean sky. You look up, the star field is dazzling. But here and there is the odd nebulous smudge of a galaxy. The great galaxy in the constellation Andromeda, for instance, is one of our near neighbors, and yet even this nearby citadel of stars is resolvable into detail only by telescope.

So that's how galaxies look to the naked eye, like tiny thumbprints in the sky. Yet, the Milky Way Galaxy holds between two hundred to four hundred billion stars, and has special significance, since it is the home galaxy of planet Earth.

Looking Down on a Galaxy from Above

Now imagine taking a galactic trip in the *Millennium Falcon*.

Given its legendary reputation on the Kessel Run hyperspace route, the outer reaches of our Milky Way will be like a walk in the park. The hyperdrive of the *Millennium Falcon* is engaged and the local starscape blurs as we fly off to the outer limits of our Galaxy.

Once in position, through the window of the *Millennium Falcon*, the

Milky Way is a fabulous, jaw-dropping sight. Looking down on our Galaxy from above, it looks like a huge, luminous spiral, consisting of around two hundred billion stars, one of which is the Sun.

From one edge to the other, the Milky Way could be as much as 150,000 light-years across. Space must be truly huge to house so many stars. So huge, in fact, that we need this special unit of light-years with which to measure it. A light-year is the distance light travels in 365 Earth days. Light is the fastest thing known to us. A beam of light will travel a little over 186,000 miles in one second. This means light can travel 5,874,601,670,040 miles in a year.

It also means that our Milky Way is 881,793,805,977,541,248 miles across!

The Geography of the Star Wars Galaxy

The Star Wars galaxy is around the same size.

It's divided into regions, with the Deep Core as the central and most luminous region of its space. The populous Core Worlds are where the human species first evolved. The Core includes planets such as Alderaan and Coruscant, which were granted permission to settle new planets. They were followed by the Colonies (which included the planets Castell and Halcyon), the Inner Rim (which housed the world of Onderon), the Expansion Region (housing the likes of Aquaris), the Mid Rim (which included Naboo), and Outer Rim Territories (which included Hoth and Tatooine).

The areas of the Star Wars galaxy were sometimes referred to as the galactic north, south, east, and west. The Unknown Regions existed to the galactic west, and remained largely unexplored throughout galactic history, as the many trade routes headed outwards towards the galactic east. Wild Space referred to regions at the very edge of the galaxy, inhabited by sentient species, but never fully charted, explored, and civilized.

Almost all of the inhabited worlds, Tatooine, Naboo, Hoth, and so on, lie in the Expansion Region of the Star Wars galaxy. Looking down from above, the Expansion Region would sit on the right of our stunning sight.

Jakku, Endor, and Coruscant are situated around the galactic core.

If this distribution of inhabited worlds were the same in our Galaxy, then our Sun and its family of planets could be sitting in the Milky Way's version of the Unknown Regions. It might even be called the dark side of the Galaxy.

The Fermi Paradox

This situation poses an interesting and possible solution to the question of life elsewhere in our Milky Way.

For generations, if not centuries, astronomers had wondered whether, in a galaxy assumed to be full of alien life, we should by now have seen some evidence. The question was neatly summed up in the Fermi Paradox, "Where is everybody?" Enrico Fermi was an Italian physicist who created the world's first nuclear reactor and received the Nobel Prize for Physics in 1938 for his discovery of transuranium elements.

Fermi's remark came while he was chatting over lunch with colleagues about the possibility that many sophisticated societies populate our Galaxy. Most of the scientists thought it safe to assume that we humans have a lot of cosmic company. However, Fermi's keen brain realized that if this were true, if there really *are* a lot of alien civilizations, then some of them would have spread out, as has been assumed since in the case of Star Wars galactic history.

So Fermi understood that any civilization with the necessary rocket technology, and the outward urge of imperial incentive, could swiftly colonize an entire galaxy. By his calculation, within ten million years, every star system could be brought under the jackboot of the Empire, or a similar tyranny. Ten million years is not as long as it may sound, compared with the age of our Galaxy, so imperial conquest of the Milky Way should be a relatively rapid affair.

Now, if these alien civilizations have had more than enough time to pepper the Galaxy with their presence, where *are* they all? Where are the legions of sleek spaceships tearing across the sky? Where are the hordes of invading aliens? Where's the future promised in sci-fi?

Solutions to the Fermi Paradox

Many researchers have considered the Fermi Paradox a radical conclusion drawn from such a simple observation: if intelligent extraterrestrials exist, humans would surely by now have had some close encounter with them. Many have offered believable solutions.

Perhaps a potentially expansionist alien civilization would self-destruct before they developed interstellar space travel? Maybe the stars are so far apart that space travel is simply too tricky and expensive? Maybe aliens *do* visit, but keep their visits and observations secret until humans have settled their petty differences?

Star Wars offers another option.

Let's assume the Sun and Earth lie on the dark side of the Galaxy, in the Milky Way's version of the Unknown Regions. This could mean our solar system is just too far from the fashionable Core Worlds of the Milky Way. It would take aliens a journey of trillions of miles, either through the dense galactic core or through its spiralled suburbs, to get to the solar system.

Maybe in the Milky Way, too, there are many trade routes headed outwards towards the galactic east, and in our galactic history the galactic west is inhabited but not yet fully explored and charted.

Maybe we are not alone after all.

DID STAR WARS PREDICT THE EXISTENCE OF EXOPLANETS?

Remember that scene in *The Force Awakens* where Han, Rey, and Finn journey into Maz Kanata's colorful dive bar in that castle on Takodana?

A haunt of freight pilots and dangerous alien pirates, the scene knowingly harks back to Tattooine's Mos Eisley cantina in *A New Hope*. Another dimly-lit tavern, known for its strong liquor, jazzy tunes, and occasional outbreaks of shocking violence, it's where Luke first meets Han in the saga opener.

But where do all those various alien races actually come from? The answer: Exoplanets. Thousands of them.

The memorable drama in the two taverns reveals that, even in the most ludic scenes, the Star Wars universe is not only replete with life from alien worlds, but also subtly philosophical. And here's why.

The Assumption of Mediocrity

The Star Wars universe is a good example of the principle of mediocrity.

The mediocrity principle says that, given the existence of life on Earth, life should also exist on Earth-like planets throughout our universe. The principle has also been used to suggest that there's nothing very unusual about the evolution of our solar system, Earth's history, the evolution of biological complexity, human evolution, or any one nation. Or, in

the rather unromantic words of British theoretical physicist, Stephen Hawking, "The human race is just a chemical scum on a moderate-sized planet, orbiting around a very average star in the outer suburb of one among a hundred billion galaxies."

With this in mind, you can see that Star Wars is in many ways a philosophical statement about the place of humanity in the universe. The franchise assumes mediocrity in its portrayal of a galaxy full of alien life-forms, whose technology is more advanced and whose powers are greater than ours. Star Wars certainly does not start from the assumption that the human phenomenon is special, privileged, exceptional, or even superior.

As with many philosophical and scientific notions, the principle of mediocrity began with the ancient Greeks. In the fifth century BC, thinker Leucippus and his pupil Democritus suggested all matter was composed of atoms.

These "atomists" also had the first truly cosmic view of our universe. They may have been the first thinkers to realize that the fuzzily luminous Milky Way would resolve into the light of discrete but distant stars. They believed in numerous other worlds, and in extraterrestrials. For the atomists, the life-bearing worlds they visualized were beyond reach. We could say, "far, far away."

Another atomist, Epicurus, suggested there were an infinite number of worlds, derived from an infinite number of atoms, beyond the senses, but not beyond reason. It's clear Epicurus believed aliens lived on these other worlds:

> *"We must believe that in all worlds there are living creatures and plants and other things we see in this world; for indeed no one could prove that in a world of one kind there might or might not have been included the kinds of seeds from which living things and plants and all the rest of the things we see are composed, and that in a world of another kind they could not have been."*

In the ancient debate, there were those—unlike the atomists—who wished to limit the heavens. They kept with the old creed, that the Sun was a planet, like the other planets in orbit about the central Earth. However, the atomists wished to open up creation in their argument. They held that the Sun was a star, the stars were all suns, and suns always came with planets. Lots of them.

The debate also revolved around whether the Earth was the center of the universe, or just a mere planet in it. If the Earth was a planet, then the planets could be Earths. If the Earth was not central, then neither is humanity. Now you can see why the Star Wars universe is decentralized, infinite, and alien.

Another atomist was the Roman poet and philosopher, Lucretius. His poem, *On the Nature of Things* (in which he sounds a little like a Jedi), states his belief in mediocrity:

> *"Turn your mind first to the animals. You will find the rule apply to the brutes that prowl the mountains, to the children of men, the voiceless scaly fish and all the forms of flying things. So you must admit that sky, Earth, Sun, Moon, sea, and the rest are not solitary, but rather numberless."*

I-Spy

For over two thousand years, from the Greeks to the Star Wars saga, such ideas held firm.

There was continuous belief in the principle of mediocrity, and the notion of life-bearing worlds out in the depths of space. However, no evidence was ever found.

Even with the revolutionary invention of the telescope in the early 1600s, the truth still lay beyond our senses. It slowly became clear that the stars lay at huge distances. To cope with this, writers used reason to try imagining the unimaginable: what worlds, and what kind of creatures, exist beyond the range of our "far-seers," as the telescope was also known. And the belief burgeoned that the stars were indeed suns.

By 1977, when the Star Wars saga started, modern sci-fi had been imagining alien worlds for almost one hundred years. Science fiction had inspired real science. Funding programs for the actual scientific search for extraterrestrial intelligence began in the 1960s. Scientists started to search for powerful radio signals from space.

Star Wars arose within this culture of science and science fiction.

A New Hope dreamt up other worlds. The desert planet of Tatooine; the mining planet of Kessel; the gas planet of Yavin, and its forest moon of Yavin 4; the rural planet of Dantooine; and the industrial planet of Corellia, Han's home world. To date, there are over fifty various exoplanets in the canon, including the latest additions, such as the ocean planet of Ahch-To and the desert planet of Jakku.

However, it was only after the saga began that actual exoplanets were first discovered.

In the year of *A New Hope*, NASA had launched the two robotic Voyager probes to the outer solar system. They have since become the first human-made objects to enter interstellar space, traveling "further than anyone, or anything, in history." On their journey to the stars, the Voyager probes discovered evidence that our own gas giant moons, such as Europa and Titan, were worlds in their own right. What other worlds lay beyond our system? What extra-solar planets, or exoplanets, may dwell there?

Worlds Beyond Our Sun

Today, we live in a great age of discovery.

We live at a time about which many writers and thinkers could only dream. The first confirmation of an exoplanet orbiting an ordinary star was made in 1995, almost twenty years after *A New Hope*. This giant planet was found in a four-day orbit around the nearby star 51 Pegasi.

Since 1995, astronomers hunting for potential life-bearing, Earth-like planets around Sun-like stars reckon there may be tens of billions in our Galaxy alone. A European team of scientists reported that perhaps 40% of the estimated 160 billion red dwarfs in the Milky Way have a "super-Earth" orbiting at a distance that would allow water to flow freely on its surface.

In a very real sense, the Star Wars universe has been reborn.

The American space observatory *Kepler*, launched in 2009 to find Earth-like planets orbiting other stars, took off four hundred years after Galileo's first use of the telescope, and is of course named after the great German theorist, Johannes Kepler. Based on *Kepler*'s early findings, Seth Shostak, senior astronomer at the SETI institute, estimated that "within a thousand light-years of Earth," there are "at least thirty thousand habitable planets."

Based on the same findings, the *Kepler* team projected that there are "at least fifty billion planets in the Milky Way," of which "at least 500 million" are habitable. NASA's Jet Propulsion Laboratory was of a similar opinion. JPL reported an expectation of two billion "Earth analogs" in our Galaxy, and noted there are around "50 billion other galaxies," potentially bearing around one sextillion Earth analog planets.

A World of Your Own

Ten years before *A New Hope*, science fiction writer Arthur C. Clarke had written about the possibility of other worlds beyond our solar system.

"Almost certainly," Clarke said, "there is enough land in the sky to give every member of the human species, back to the first ape-man, his own private, world-sized heaven—or hell. How many of those potential heavens and hells are now inhabited, and by what manner of creatures, we have no way of guessing."

Yet, Star Wars has helped us guess. The saga has helped us realize that the barriers of distance are crumbling, as we continue to venture further into space. Star Wars was not the first to imagine alien worlds, but it has conjured up a compelling portrayal of what an inhabited galaxy might look like.

The truth, as always, will be far stranger.

WHAT DO STAR WARS EXOMOONS MEAN FOR LIFE IN OUR GALAXY?

In *The Force Awakens*, we meet two new exomoons: the twin moons of the planet Jakku. All three are orbiting a single sun.

Although the moons of Naboo were also habitable and at one time hosted spice mines, to date, the most famous exomoon in the Star Wars saga has arguably been Endor. For one thing, it's certainly been the most dramatically important. The Battle of Endor saw the revolution to restore the Republic. It was here that the Imperial forces kept their shield generator, which at first prevented the Alliance from destroying the Death Star.

The forest moon of Endor, according to canon, was about 4,900 kilometers (3,000 miles) across, which is easily bigger than our Moon. Endor was roughly 43,000 light-years from the galactic center, covered in dense woodlands, and had a breathable atmosphere. Endor had two suns, Endor I and Endor II.

The moons of Jakku, Naboo, and Endor are part of a growing catalog of Star Wars exomoons. But have we found anything similar in our exploration of space? If so, what does it mean for life in space?

Rey the Stargazer

Let's imagine Rey as a stargazer.

As a scavenger left behind on Jakku when she was a child, she may have had plenty of time to gaze at the stars, weather permitting. Now, the view of her night sky wouldn't be much like ours. True, she can see two moons, rather than our one. Living in the Western Reaches of the Inner Rim, her planet is closer to the core of its galaxy than Earth is to the Milky Way's. There are thousands of stars visible to her naked eye, and billions beyond her sight. However, outside of the immediate Jakku system, none of the other planets and moons of the Star Wars galaxy can be seen by Rey.

That's because exoplanets are very faint points of light, compared to their suns. In fact, they're usually one million times fainter than their parent stars. And exomoons would be even fainter. Rey has no chance of seeing them. Looking for them would be like looking for a needle in a haystack (assuming they *have* haystacks on Jakku). However, the scientists of the Core Worlds would have known of the existence of the many exoplanets and exomoons in their galaxy for many generations. So how would they have spotted them?

Planet Hunting with the Ancient Coruscanti

Now imagine alien planet hunters, sitting on ancient Coruscant.

This is the world that was to become the center of political and cultural life of the Star Wars galaxy. And this is no mere stargazing. If ancient Coruscanti tracked the movement of neighboring stars wheeling a way through their neighborhood of the galaxy, they may have seen some of them wobble a little.

Although the stars are huge, they're not as strong and stable as they seem. A star like Alderaan with a satellite planet (also called Alderaan) will move in its own wobbly orbit, due to the planet's gravity. So when the Coruscanti planet-hunters see a wobbly star, they know there may be a planet in tow. Also, by measuring the size of the wobble, the Coruscanti can approximate how big the planet is.

In the early days, looking for wobbly stars may have been the best way for the ancient Coruscanti to find exoplanets. At least, that's how it's done on Earth. Thousands have been discovered in our neighborhood of the Galaxy. Planet hunters look for twitchy stars out to about 160 light-years from Earth. And, like all stars with planets, our Sun wobbles, too. The combined gravity of the planets is what causes the Sun's wobble, especially mighty Jupiter, which is bigger than the other planets put together.

The "Jupiter effect" means the Coruscanti may find a very large planet in orbit around a wobbling star. Because bigger planets means bigger wobbles, this way of planet hunting is best for finding other Jupiter-like gas giants such as Bespin, which is about 85% of Jupiter's diameter.

Perhaps the Coruscanti called it the "Bespin effect," although that's highly unlikely, as the Bespin system sits in Outer Rim territories and is probably impossible to "see" from Coruscant, which is on the opposite side of the galaxy's dense galactic core. They're also unlikely to have used the wobble method for finding Alderaan, too. As a terrestrial planet, Alderaan is not big enough to make a measurable wobble in its parent star. For smaller planets like Earth and Alderaan, you need another method.

So Coruscanti planet hunters could also have used eclipses to find exoplanets.

In space, an eclipse happens when one body passes into the shadow of another. So, an eclipse of the Sun happens when it moves into the shadow of the Moon, and the Sun cannot be seen from Earth. In other star systems, a planet might cross in front of the parent star, causing a mini-eclipse. If this happens, hunters might find an exoplanet in that system—and where there are planets, there are also moons.

The Goldilocks Zone and the Red Dwarf

The ancient Coruscanti would also be interested in Goldilocks Zones.

In planetary systems, most scientists believe that a planet must lie in the habitable zone to support life. This is the range of various orbital distances around a star within which rocky planets can support liquid water on their surfaces. If a planet is beyond the outer limits of the habitable zone, it will not get enough of the Sun's energy, and water will freeze. If a planet is within the inner limits of the habitable zone, it will be exposed to too much solar energy, and surface water will boil away. This not-too-hot, not-too-cold idea is why scientists sometimes call the habitable zone the Goldilocks Zone.

Looking for life isn't just a simple matter of looking for exoplanets. The Coruscanti may have narrowed their aim to the Goldilocks Zones.

And here, red dwarf stars become important in our own hunt for exomoons.

Our Milky Way Galaxy contains between two hundred and four hundred billion stars, although the actual number may be as high as one trillion. Around the orbits of these stars exist at least one hundred billion planets, many of them Earth-like.

Now, according to estimates, red dwarfs make up 75-80% of the stars in the Milky Way. They are by far the most common type of star. The same can also be assumed of the Star Wars galaxy. But that proves an interesting problem for life on planets in orbit about red dwarfs.

A red dwarf is a small, cool star. Such stars have masses less than half of the Sun's. Red dwarf stars were once thought unable to support habitable planets. However, scientists now think that old theory is wrong. Red dwarfs *do* have planets, but they have curious orbits. Planets in the Goldilocks Zone of a red dwarf would be so close to the parent star that they would be tidally locked. Just like the Moon in orbit about the Earth, the red dwarf's planets always show the same face to their sun.

So, red-dwarf worlds are locked by the parent star's gravity. One half of the planet is always in darkness, the other half always bathed in light. They are strange and exotic worlds. On the dark side, there is a vast frozen waste. On the light side, oceans, a temperate climate, and land. However, there is also a "Twilight Zone," an in-between place. Here, between the contrasting half-worlds, strange creatures may compete for food and light—if there are any creatures at all in such an alien world.

But even if a red dwarf exoplanet doesn't harbor life, its exomoon may.

Imagine an Earth-sized moon, in orbit around a Jupiter-sized giant. The giant may be locked in its orbit around the red dwarf, suffering extremes of climate. Yet, the moon in orbit around the giant may be habitable. It would circumvent the tidal lock problem by becoming tidally locked to its planet. This way there would be a day/night cycle as the moon orbited its primary, and there would be distribution of heat.

Many Endors?

This is why exomoons may be so important.

Since the rather humble red dwarfs predominate, then the planets and moons in orbit about them will also do so. As red dwarfs make up three-quarters of the stars in the Milky Way, their planets and moons may even prove to be the norm. And, as the planets themselves may be inhospitable half-worlds, their moons may be a common home of life in the Galaxy.

Consider the lifetimes of red dwarfs. They are huge. A red dwarf half the mass of the Sun has a lifetime of fifty-six billion years. At the moment, it's not clear whether the energy output of red dwarfs is stable enough for the development of life. But if civilizations do develop on red dwarf exomoons, they could have huge lifetimes, too. It's possible that the meek truly will inherit the Galaxy.

Could Endor be the greatest of all Star Wars insights? Endor is a habitable forest moon in orbit around an inhospitable gas giant. The diameter of Endor is about 40% of the Earth's. How common are such worlds in our Milky Way? The answer, as of right now, is that we don't know.

Maybe tomorrow, or maybe a decade or century from now, we may make the most shattering discovery of all time: the discovery of a thriving extraterrestrial civilization. As the twenty-first century dawned, we had been imagining alien life for almost two and a half millennia. With the current state of technology, we can only wonder at the chances that civilization lives on an exomoon.

COULD STARKILLER BASE DRAIN THE ENERGY FROM A STAR, AS IN THE FORCE AWAKENS?

Starkiller Base. Function: headquarters of the First Order junta. History: built after the blasting of the second Death Star in the skies above Endor. Technical brief: a transformed ice planet from the Unknown Regions, Starkiller Base was more than twice the size of previous battle stations, and significantly more powerful. Think planet Pluto with a powerful pistol. In short: *Death Star III*.

Just in case there's any doubt about the might of this later model of killer space station, the Base proved its power early on in *The Force Awakens*. It annihilated *five* planets at once. As the name Starkiller implies, the Base was powered by draining energy from burning stars. How might it have done that? And how would it have stored the energy once a star had been drained?

Atom-Smashing Machines in the Sky

Starkiller Base was not the only atom-smashing machine. The universe is full of them. They're better known as stars.

Stars power our Galaxy, as they would the Star Wars galaxy. Stars are bringers of energy and light, the building blocks from which the entire cosmos was created. Without our own star there'd be no light, no life, no Earth as we know it.

Stars such as the Sun are huge balls of burning gas. It's mostly hydrogen, but about a quarter of it is helium. Their composition depends on how old they are. More highly evolved stars will have a greater amount of gases heavier than hydrogen and helium.

The architects of Starkiller Base would have been well aware that stars like the Sun have core temperatures of almost sixteen million degrees. That's hot enough for nuclear fusion, such as turning hydrogen into helium. Atom-smashing happens in all stars, due to these huge temperatures and pressures at their very cores. To be clear, it's more like atom-fusing than atom-smashing, but the gassy constituents have to be smashed together before they can fuse.

Starkiller Base as an Atom-Smasher

Atom-smashing liberates huge amounts of energy that could have been tapped by the Starkiller Base.

Stars like the Sun burn about four million tons of gas, every second. That's as much energy as seven trillion nuclear explosions, per second. Yes, that's right: when you sit in the Sun, you're basking in the glow of seven trillion nuclear explosions a second. It's certainly enough energy to kill a planet or five.

In fact, let's use Einstein's famous equation $E=mc^2$ to work out how much energy you could get from fusing all the Sun's hydrogen into helium. Plugging in the figures for the mass of the Sun (that's the "m" in the equation) and the speed of light (the "c"), we get an answer of 870 million trillion trillion trillion Joules of energy. That's gargantuan. Draining the Sun alone would provide enough energy to zap four trillion planets, every planet in the Star Wars galaxy.

The Starkiller architects would have known stars are *stable* sources of energy, too. Stars of the Sun's mass will burn hydrogen for about ten billion years. There is a huge range of stellar sizes and masses, of course. In fact, stars of smaller mass hang around even longer than stars like the Sun. Small stars have what are called "forever lifetimes." A star half the size of our Sun will go on burning for fifty billion years.

The Fifth Element

So, stellar energy sources are long established. How could Starkiller Base have tapped that immense energy? According to canon, the Starkiller weapon ran on a form of dark energy known as "quintessence," which permeated all corners of the universe. It's time for a quick quintessence backstory.

The concept of quintessence is ancient. It came from the Greek thinker, Aristotle. Back in the days of classical Greece, philosophers reasoned that all matter was made up of different combinations of the four elements of earth, air, fire, and water.

It's an ancient mega-simple style of the Periodic Table of Elements, except there are only four elements in the list. Each element had a natural place to be, earth downward, fire upward, air and water horizontally. These four elements were agents of change. They continually transmuted, and Earth was replete with their fusions.

The heavens, on the other hand, were made of the quintessence, the fifth element. Chaste and immutable, some ancient Greek thinkers considered the cosmos, in orbit about the central Earth, to be a crystalline form of quintessence. And if you flew far enough, out into the depths of the heavens, the purer the quintessence became, until it met its purest form in the realm of God, the prime mover of the universe.

Does Star Wars quintessence have a ring of truth? Yes. Sort of. Star Wars quintessence is described in canon as being ubiquitous in the universe. How does that compare with our universe?

From what is currently *hypothesized* about dark energy, there are two forms. The first form, the cosmological constant, uniformly permeates all of space. There'd be no point looking specifically to the stars for this first form of dark energy. Real scientists know the second form of dark energy as quintessence. That's because it, allegedly, represents the fifth "known" contribution to the total mass-energy content of the universe, along with baryonic matter, radiation, cold dark matter, and gravitational self-energy. Unlike the cosmological constant, real quintessence doesn't uniformly permeate all of space, but is a scalar field, which means varying values at every point in space. More common examples of scalar fields include temperature and pressure.

How Starkiller Base Is Meant to Work

To power the quintessence weapon of the Starkiller Base, the First Order looked to the stars for unlimited energy. In short, the Base drained stars to charge its big gun.

According to canon, the Starkiller Base worked by focusing on a single star as a power source. Then, using an array of collectors on one side of the planet, this array would gather "quintessence" in phases, directing it down to the planet's core. Here, it was held in place by the planetary magnetic field, augmented by an artificial containment field, installed within the planet's crust, and maintained by machinery at the First Order's disposal.

You can see that canon gets very confused with itself. It's unclear whether the Base used real stellar baryonic matter ("starstuff") to charge its gun, or dark energy (quintessence) to do so.

In some instances Starkiller Base is described as draining routine starstuff to power its superweapon. Other accounts not only have the Base draining dark energy instead of ordinary matter, but go on to explain that the superweapon "fired dark energy" (this is the dark side after all), which is transformed on firing into a state known as "phantom energy" (phantom being another favored word). Canon can't make up its mind.

Starstuff Works Best

For the sake of argument, let's take sides and assume the Base used actual starstuff to power its weapon. The brief calculations we did above clearly show there's enough energy in the Sun alone to atom-smash those planets, without having to invoke anything as exotic and elusive as "dark energy."

After all, when Starkiller was in operation, it had all the appearances of what's thought to happen when a star gets too close to a black hole—a vortex of heat and light, spiralling into gravitational oblivion. The sight has all the hallmarks of ordinary starstuff, draining down into the Base. If this were the case, we could assume that the First Order had worked out a way to deal with the heightened gravity of the Base by somehow harnessing an engine that could fuse the star fuel for its quintessence superweapon, once the fuel had been drained.

As Starkiller Base charged up from stellar power, it gradually blocked out starlight until, running at full facility, it extinguished the star completely, leaving the surface in darkness. When the quintessence superweapon fired, the Base engineers would induce a breach in the containment field, allowing the collected energy to escape the core through the hollow cylinder opening on the opposite side of the Base relative to the stellar collector.

Curtains for Starkiller Base

Maybe the destruction of Starkiller Base can settle the score.

The Base was ultimately destroyed by exploiting a weak point in an attack led by pilot and spearhead of the Resistance, Poe Dameron. The strike caused the implosion of the entire Starkiller Base planet, thirty seconds before the quintessence superweapon fired upon the Resistance base on D'Qar. The canon account suggests that "as it was destroyed, the stored material from the sun it drained expanded to create a new star in the planet's place, turning the star system into a binary."

This so-called "material from the sun" sounds much more like ordinary starstuff than dark energy. At least that puts the function of the weapon of mass destruction into the realms of the merely improbable. Just exactly how Starkiller drains down all that mass without morphing into a black hole is another matter.

WHICH EXOPLANETS ARE LIKE STAR WARS WORLDS?

Jakku. Tatooine. Hoth. Planet names familiar to aficionados of that galaxy far, far away. But what of Kepler-16b, 51 Pegasi b, or even OGLE-2005-BLG-390Lb? They're the given names of exoplanets found orbiting other stars in our Milky Way Galaxy.

Sure, the Star Wars saga got there first. It helped conjure up a compelling picture of what inhabited planets might look and sound like. But now scientists are finding that our stellar neighborhood is filled with exoplanets. Many of them are as exotic as anything in the saga. Some even bear more than a passing resemblance to their fictional counterparts.

Making Sense of the Milky Way

Before wading through the weird and wonderful worlds, which have been found out in deep space, it's useful to start the journey in our own backyard. After all, our solar system acts as a neat yardstick against which other systems can be gauged.

In our system, there are two kinds of planets: rocky and gassy. The basic difference is this: On rocky planets, there's somewhere to land the *Millennium Falcon*—the ground. However, gassy planets don't have "grounds"; they just have gas. This makes landing more difficult. Rocky planets form close to the Sun, in the vicinity of the Goldilocks Zone. The bigger gassy planets form further out, beyond the frost line, and way into the cold zone.

Gassy giants have been found in other systems, too.

As planet-hunters scan the skies for wobbling stars, it was known in advance that bigger planets would mean bigger wobbles. As massive giants, they simply make larger and more rapid wobbles in their parent star, making their presence easier to detect.

That's why the vast majority of known exoplanets are so-called "hot Jupiters," or perhaps "boiling Bespins," using Star Wars vernacular.

These hot Jupiters (sometimes also called "roaster planets") are so-named as their mass is of Jupiter proportions, but their orbit is much closer to their parent star than Jupiter's is to the Sun. Between ten and three hundred times closer, in fact. Being so close to the mother star means that hot Jupiters get their gassy atmospheres stripped away by the heat. So this class of exoplanet is quite unlike the liveable worlds in Star Wars. And yet, as detection methods got better, scientists began to find exoplanets far more in the Star Wars style.

Consider Kepler-16b

Now consider NASA's *Kepler* Mission. Launched in 2009, the job of the *Kepler* spacecraft is to spot Earth-like planets among the 155,000 stars in the constellations of Cygnus and Lyra, which are visible from our northern hemisphere.

To date, *Kepler* is the most exciting exoplanet search mission. The *Kepler* space observatory looks for Earth-like planets orbiting Sun-like stars by scanning the sky, looking for mini-eclipses, those tell-tale signs that an exoplanet is crossing in front of its parent star. By using Cygnus and Lyra as examples, the *Kepler* mission also aims to estimate how many of the billions of stars in our Galaxy may have Earth-like worlds.

What *Kepler* found is quite incredible.

When the *Kepler* observatory completed its primary mission objectives in 2012, it had detected nearly five thousand exoplanets. Perhaps its strangest discovery is a so-called Styrofoam planet, a world with just one-tenth the density of Jupiter. And its most stunning discovery is the first confirmation of a rocky planet outside our solar system. In 2015, *Kepler* scientists announced the discovery of the Earth's "closest twin."

However, *Kepler* had also found a Tatooine-like planet in Kepler-16b.

Like Tatooine, Kepler-16b enjoys double sunsets, as it circles two stars rather than a single sun. Scientists confirmed this first unambiguous discovery of a circumbinary planet in 2011, a full thirty-four years after its fictional counterpart was dreamt up in *A New Hope*.

Kepler-16b sits in a system about two hundred light-years from Earth. Its mass is similar to Saturn, which makes it much bigger than Tatooine, whose diameter is only 82% of the Earth's. The Kepler-16 stars are small compared to our Sun, at about 69% and 20% of the Sun's mass. Tatooine's binary stars are both roughly Sun-sized.

The story of Kepler-16b is a good example of how planet-hunting works. Pictures captured by Kepler's camera showed two stars orbiting each other. Next, eclipses were seen, where one star moved in front of the other. However, with a closer inspection, the evidence showed subsequent eclipses that couldn't be explained by the movement of the two stars alone. Instead, the tiny drop in light from the stars—a dimming of only 1.7%—was found to be the tell-tale sign of an orbiting planet.

Conditions on Kepler-16b

Kepler-16b is far cooler than Tatooine.

Down on the surface of this giant, frosty world, a planetary explorer would experience a temperature range from -70°C to -100°C. Chilly Kepler-16b would never see constant daylight as the two stars are too close together. The binaries would come together in an eclipse every 20.5 days, then move apart again. As their separation in the sky increased, they would go down at different times, in the kind of sunsets never spied on Earth but often seen on Tatooine.

Could Kepler-16b support life?

It certainly orbits within the Goldilocks Zone. The Zone of the Kepler-16 system runs out from about 55 million to 106 million kilometers away from the binary suns. Kepler-16b lies at an orbit of 104 million kilometers, so it sits near the outer limits of its habitable zone. However, it's worth noting that this is a gas giant with freezing temperatures.

So chances of life on Kepler-16b appear remote. But what about an exomoon? Sometime in its history, Kepler-16b could have captured an Earth-sized world from the center of its habitable zone. Perturbations from other bodies in the Kepler-16 system could have caused this "Earth" to migrate, sending it on a trajectory that ended up with Kepler-16b making the "Earth" its moon.

Life could also develop outside the Kepler-16 Zone. Research shows that it's also possible for this system to support a faraway habitable planet orbiting at about 140 million kilometers away from the center. If this planet had a thick enough mix of greenhouse gases in its atmosphere, including carbon dioxide and methane, it could retain the heat needed to keep water in liquid form on its surface.

Circumbinary Sweet Spots

All these possibilities are fascinating and exciting. It's worth remembering that, for many years, astronomers said that planets could not form around binary stars, due to the effects of gravity in such systems.

The Star Wars saga ignored all that.

Tatooine certainly sits in the circumbinary sweet spot of the Tatoo system. Not only has Kepler-16b been discovered, but new research also shows that far from being only merely possible, Tatooine-type planets may actually be quite common. It seems there might also be similar sweet spots in other systems.

Recent research has shown that in the region near a binary star, a rocky or gas-giant planet can form in much the same way as around a single star. In other words, planets are just as prevalent around binaries as around single suns. The conclusion? Tatooines may be common in the universe.

The Case of OGLE-2005-BLG-390Lb

And that brings us to another exoplanet, discovered close to the center of our Milky Way Galaxy.

This Inner Rim world, as it justifiably could be called, is an Earth-like planet that circles its parent star about every ten years. The planet, OGLE-2005-BLG-390Lb, orbits a red star five times less massive than the Sun, and is located at a distance of about twenty thousand light-years from us.

However, OGLE-2005-BLG-390Lb sounds more like Hoth than it does Earth.

From space, Hoth looks like a pale blue orb, due to its cover of dense ice and snow. Hoth is the sixth planet in its system, which means the temperature, although always freezing, can drop to -60°C at night.

OGLE-2005-BLG-390Lb sounds similar. With a cool parent star and large orbit, this exoplanet is likely to have a surface temperature of -220°C, too cold for liquid water. It's also likely to have a thin atmosphere, and a rocky surface buried under layers of ice or beneath frozen oceans. OGLE-2005-BLG-390Lb is about five times the mass of the Earth, whereas Hoth has a diameter similar to that of Mars. Indeed, in its system, OGLE-2005-BLG-390Lb sits at an average distance of 2.0 to 4.1 AU (or Astronomical Units—the Earth–Sun distance). This means that in our solar system, OGLE-2005-BLG-390Lb's orbit would fall somewhere between that of Mars and Jupiter.

Where Your Shadow Always Has Company

NASA marketing was swift in realizing the importance of exoplanets.

In 2015, the US space agency published a series of striking travel posters, created by their visual strategists Joby Harris, David Delgado, and Dan Goods. Using Art Deco typography, bold colors, and a classic design—evocative of the golden age of travel—the posters invite tourists to visit recently discovered exoplanets. The posters included one on dual-star planet Kepler-16b, which boasts the strapline: "Where your shadow always has company."

David Delgado explained how they had felt inspired by the discovery of so many new exoplanets: "We thought it would be really cool to explore the characteristics of each planet through the context of travel," he said. "It feels like we're living in the future, or science fiction is coming to life."

THE LINK BETWEEN HOTH AND MARS: WHY MIGHT HUMANS MAKE LIKE EXOGORTHS?

The pirate starship sits in a dark, desolate asteroid cave.

The *Millennium Falcon* cockpit is silent, lit only by the pulsing hyperdrive lights of the flight instruments. Leia sits alone in the pilot seat. Some movement outside the cockpit window catches her eye. With the reflection of the panel lights, it's hard to make out what it is. She moves closer to the glass and gazes into the gloom.

The dark is so bible-black, the sheer extent of the cave is impossible to fathom.

For how far does the cave system run?

In science fiction movies such as *The Empire Strikes Back*, extraterrestrial caves are cosmic catacombs that can even accommodate spaceships in flight. On planet Earth, there are thousands of miles of unexplored caves. But what of the caves on the Moon, and planets such as Mars? How big could they be? How long are they? We know little of them. Might this dark side of the solar system be used to help colonize space?

The Moon as a Stepping Stone

The Moon has been a target of human colonization for many years. It's on our cosmic doorstep, within easy reach of Earth.

In the Star Wars galaxy, many of the species that would become predominant originated in the Core Worlds, including humans. Although conflicting theories on human origins abounded, prominent historians believed that Coruscant was their most likely homeworld. The old Coruscanti would have faced a similar kind of colony challenge in the very early days of the exploration of Star Wars space.

The closeness of our satellite might enable colonists, living and working on the Moon, to exchange goods for the things they need from Earth. It would be the beginning of our own interplanetary trade routes. The Moon is also a good place to set up powerful telescopes, and a stepping-stone for further exploration of our Galaxy.

The NASA Apollo mission program of 1961–1972 was pioneering work. The missions put a total of twelve humans on the Moon, and tested the challenges of lunar living. Each mission was longer than the previous one, to explore the Moon further and see how well the astronauts could cope while conducting different types of experiments. The Apollo program also taught us how to transport large items of equipment and materials to another world.

Make Like an Exogorth

During the development of a colony on the Moon, it may be necessary to adopt exogorth tactics and dwell in convenient caves.

Visiting the Moon is one thing, but living there won't be easy. There are huge temperature swings, from 134°C (273°F) at noon to -170°C (-274°F) at night. The surface is constantly blasted by micrometeorites and harmful cosmic rays. To survive this bombardment and radiation, colonists might have to live underground in lunar "lava tubes."

Lunar lava tubes are subterranean tunnels on the Moon, which are thought to have formed by the flow of magma in the ancient past. When the surface of a live lava tube cooled, it formed a hardened lid that contained the on-going lava flow under the lunar surface, in tunnel-shaped passages. Once the flow of lava died down, the tunnels would become drained, forming hollow catacombs.

The exogorths got it right. Lava tubes are a very useful living space. Many are long, natural caves, as wide as five hundred meters (1,600 foot) before they become prone to gravitational collapse. Even if the tubes survive, shifting seismic events—or meteoroid bombardment—may still disrupt stable passages. No wonder Leia was anxious about the strange, creaking noises coming from the asteroid catacombs at night.

Naturally, not all lunar living would clone and copy the ways of the giant slug.

Mynocks are right off the menu on the Moon. Indeed, the availability of food and water is a major challenge for lunar colonization. At first, food would need to be brought along—more than just a packed lunch, of course. However, scientists believe that water is hidden in the soil at the Southern lunar pole. Special machines would be needed to extract the water from the ground. Growing plants will also be difficult at first. The nights can be long and cold, and the daylight bright and harsh, as there is no atmosphere on the Moon to reduce and scatter the Sun's radiation. There are also no insects there to pollinate the flowers, so an entirely new way of growing food would need to be found. These are all good reasons to dwell in the tunnels until more sophisticated living conditions can be built.

Like the ancient Coruscanti, we humans have an outward urge to explore space.

A number of nations hope to put us back on the Moon. One plan is to put a farm at its northern pole. A farm here would receive eight hours of sunlight each day, throughout the lunar summer. The farm could have a special covering, providing protection—from the fierce solar radiation—and insects for pollination. But even so, a farm measuring one hundred meters by one hundred meters would not feed one hundred people.

Missions to Mars

If the Moon were the first stepping-stone for further exploration of our Galaxy, Mars would be next.

We know more about Mars—the fourth out from the Sun—than any planet other than Earth. That's because it's the closest planet whose surface we can see through our telescopes. Where our other closest neighbor, Venus, is shrouded in a veil of mysterious cloud, the Red Planet is plain to see. Like Earth, Mars has a twenty-four-hour day, it has seasons, and it has polar caps. If there's a planet we would colonize first, it would be Mars.

The Coruscant system was ripe for colonization.

As well as the planet Coruscant itself, there were eleven planets in orbit about their parent star, Coruscant Prime. Before those planets could be used as stepping-stones into Star Wars space, Coruscant also had four moons, which could have acted as viable targets for early colonization.

For us, Mars has huge potential, so some scientists believe pioneer colonists should skip the Moon and head straight for Mars. Although Mars has much in common with Earth, it lacks some of the crucial things we need to live, such as warm temperatures and liquid water. Not only that, but a Martian colony would have to fend off global dust-storms, solar radiation, and would have to melt the polar ice into a sea twelve meters deep, covering much of the planet. All are huge tasks to undertake.

So how would pioneers go about transforming Mars and meet these crucial challenges? Once more, it's easy to see the attraction in using lava tubes until the Martian colony can truly take off. Is there a maze of Martian tunnels fit for the task?

The Tell-tale Sign of the Tubes

In 2007, NASA's *Mars Odyssey* spacecraft made a stunning discovery.

Odyssey spied entrances to possible caves on the slopes of a Martian volcano. The existence of a lava tube is sometimes revealed by the presence of such tell-tale "skylights," a place in which the roof of the tube has collapsed, leaving a circular hole in the surface. The find fueled interest in potential underground habitats and sparked searches for caverns elsewhere on the Red Planet.

Mars has huge volcanoes, one order of magnitude bigger than on Earth. The famous Martian volcano Olympus Mons reaches the impressive height of twenty-one kilometers, while the whole region of the Tharsis volcanic bulge covers a surface of over twenty-five million square kilometers.

As gravity on Mars is about 38% of Earth's, Martian lava tubes are expected to be much larger in comparison, although unlikely to house lurking exogorths.

Earthlings Become Martians

So, parts of the solar system are ripe for colonization.

Like the old Coruscanti we should hop from planet to planet, with Mars as our first home away from home. The first launch could carry an unmanned Earth Return Vehicle (ERV) to Mars. The ERV would contain a nuclear reactor, which would power a unit to make fuel, using material found in the Martian atmosphere. Two years later, a manned mission would touch down near the ERV. The crew would stay for eighteen months, exploring the planet until returning to Earth using ERV-made fuel. The crew would be replaced by another team, and a string of bases would be set up.

The Martian lava tubes would act as a main base.

The tubes may contain trapped water essential for life, and may house reservoirs of ancient ice, since cool air can pool in lava tubes and temperatures remain stable. The ability to tap into these reservoirs may provide dramatic insight into the history of possible life on Mars.

Through making like an exogorth, and using the tubes as a home, in good time Mars will be terraformed. After several decades, the Red Planet will look as blue and watery as Earth. Within a century, it could be terraformed into an oxygen-rich environment, supporting a human colony, some of whom may dream of traveling to the remote corners of the solar system, and beyond.

COULD LIFE ARISE ON A DESERT PLANET LIKE TATOOINE?

Tatooine is a sun-scorched planet that orbits two suns. Its location is forty-three thousand light-years from the Star Wars galaxy's core.

Despite its hot and arid climate, Tatooine is host to a number of different life-forms. Some have taken up residence there like the Hutts, Humans, and Rodians. However, it also has native species in the form of Jawas and Tusken Raiders.

Bearing in mind the serious lack of plant life on Tatooine (except for the black melons that grow on the Jundland Wastes), and the scarcity of water, could life arise on a desert planet like Tatooine?

Life on Earth

At the moment the only place that we know life definitely exists in the universe is on planet Earth. As such, all of our studies of life are based on what we have found in our own backyard. This knowledge has enabled us to infer the constraints on life and explore the possibilities of whether it may or may not exist elsewhere in the universe.

In addition to water, we know that all known life is based on carbon. Carbon forms a wider variety of bonds than any other element. This allows a multitude of different molecules to be made. Carbon forms the backbone of all organic molecules, which are essential for life. Other important elements of life are hydrogen, oxygen, nitrogen, phosphorous, and sulfur.

Somewhere amongst all of these chemicals, life managed to emerge. This process of life emerging from non-living matter is called abiogenesis. Although it's still not known how that happened exactly, there are some things that do seem evident.

The life-forms that were around first appear to have been tiny bacteria and Archaea. They were also hyperthermophiles, which means they had the ability to live and thrive in hot water environments.

If life on Tatooine had a similar beginning to life on Earth, it may have been hot water-based microbes, too.

Tatooine

Maybe one of the most iconic images of Tatooine is of its two suns in the desert sky.

The strength and type of radiation reaching a planet from its parent star is critical to the formation of a planet's environments and potential life. The level of solar radiation could be the difference between having an ocean planet, a snowball planet, or a desert planet.

In 2011, astronomers discovered Kepler-16b, which they nicknamed Tatooine because it orbits two stars. Since then, they've discovered many more so-called circumbinary planets. Yet, unlike these mainly gas giants, the fictional Tatooine is a rocky desert planet with no apparent surface water.

The habitable zone is the distance from a star where water can exist as a liquid on a planet. Some scientists believe that a desert planet may provide a good model for an Earth-like habitable planet. Their research showed that the habitable zone for a land planet may be three times bigger than an aquatic planet. This is because the planet can absorb more of the heat that would otherwise be reflected away by snow and ice, making it habitable further out from its star. However, its dryer atmosphere would also trap less heat than a moisture-ridden atmosphere, which would extend the habitable zone closer to its star as well.

This may not be true of Tatooine, though. A major occupation there is moisture farming, where water is extracted from the atmosphere. This means it may have more atmospheric vapor than typical desert planet models.

There is a desert-like planet that does have similarities to Tatooine though, and that's Mars. Even if it is a rather cold example.

Mars has a store of water that is not visible as a liquid on the surface, as well as a small amount of water vapor in its atmosphere. Tatooine has extremely dangerous sandstorms just like Mars. Its sandstorms blow unchecked by surface features such as plants and large water expanses. However, unlike Tatooine, water can be detected on Mars as a constituent of the polar ice caps.

Somehow Tatooine has moisture farms but no polar ice caps, possibly because it doesn't get cold enough or maybe because the moisture levels are too low. Yet still, it has many features that suggest that water did flow there at some point in its history. So what's the deal?

Water Features

As far as we know, life can't survive without water.

The driest place on Earth is the Atacama Desert, but even there life has found a way to cling on.

Armando Azua-Bustos, from the Blue Marble Space Institute of Science, has studied the presence of life in the Atacama Desert. He says that "life in the Atacama has evolved to use fog and dew as a water source. . . . Inland, life has evolved to use water that's bound to hygroscopic minerals, like gypsum and sodium chloride."

So the little moisture on Tatooine could support microbial life, but what about in its past?

The surface features on Tatooine indicate that it was once covered in enough liquid to create caves, ridges, and canyons. Judging by the depth of the canyons there must have been a long history of running water, especially when our own Grand Canyon took many millions of years to form.

Understanding that, it's likely that life first emerged on Tatooine back when there was a plentiful supply of water, given the importance of liquid water to life. Then, after a drastic change in the environment, life held on in whichever way it could, accounting for the presence of only a few plant and animal species compared to the millions we have on Earth.

Deserts such as the Eastern Sahara have been in their current condition for only about 5,000 years. Before then, up to about 10,000 years ago, there was a wet period that supported many plants and animals. Certain species still survive in and around these environments, evolving to take best advantage of what little moisture there is.

Azua-Bustos says: "In the case of the Atacama, life is able to survive as it is very efficient in finding, capturing, conserving and using water. However, we know only a little on how life is able to find and capture water, and nothing on how it's able to conserve and use it yet."

So where has all the water gone on Tatooine?

Space Rays and Water Loss

Research has indicated that there may have been a primitive ocean on Mars that held more water than Earth's Arctic ocean. It was enough water to hypothetically cover the entire planet to a depth of 140 meters. Since then, more than 87% of it has escaped into space. If we could surround Mars with all its current theorized water, it would only be dozens of meters deep.

Mars is thought to have lost its water due to bombardment from solar winds, which stripped away its atmosphere. Earth has avoided this fate due to the presence of a large magnetic field called the magnetosphere. This field is much weaker on Mars partly due its smaller size and not being able to sustain the mechanisms for a strong magnetosphere. Over time it's been speculated that Mars's field got weaker due to changes in the activity of its iron core.

Having two suns means that Tatooine's atmosphere has an increased threat of degradation from solar winds. However, Tatooine is much larger than Mars, so its magnetosphere could be stronger. Considering Tatooine appears to have an identical gravity to Earth, we can work out its mass and then its average density.

The force felt on the surface of a planet due to gravity can be found by using Newton's law of universal gravitation. This force is the same as the weight a person would experience when on the planet's surface. The equation can be rearranged to find the mass of Tatooine by using a surface gravity equal to Earth (9.81 meters per second^2) and a radius equal to Tatooine, which is 5232.5 kilometers.

Once we have the mass, we can divide it by Tatooine's volume ($4/3\ \pi r^3$) to get an average density for the planet.

Once it's been calculated, it turns out that it's about one-fifth denser than Earth. This indicates a bigger proportion of iron in its core.

As such, in Tatooine's past the extra iron could have lead to a stronger magnetosphere than Earth's. This could offset any extra attack on its atmosphere from its two suns, making it safer for life to emerge. Without adequate protection, all life on Tatooine would experience possible DNA damage from solar winds and cosmic rays.

There's also the threat of ultraviolet (UV) radiation. On Earth, life initially avoided the perils of UV light by inhabiting underwater environments. The water absorbs harmful UV rays, allowing life-forms to thrive without experiencing serious cell damage. Unfortunately, there is no sign of a protective underwater environment on Tatooine. Again, a presence of surface water there in the past would have helped.

So how else can life be protected from UV rays on Tatooine?

Oxygen and the Emergence of Life on Tatooine?

Although there are life-forms that can get by without oxygen, these are all tiny microbes. For the majority of known complex life-forms to exist, oxygen is required.

On Earth, the oxygen in the atmosphere was and is mainly produced by bacteria and plant life, through the process of photosynthesis.

Photosynthesis is how plants get their energy. They take in carbon dioxide (CO_2) and strip off the carbon to combine it with water (H_2O). This produces carbohydrates such as glucose. The remaining oxygen (O_2) liberated from the CO_2 escapes away into the atmosphere.

It's this escaping oxygen that eventually led to the formation of the ozone layer, which protects land-based life from the Sun's UV rays. The presence of the ozone layer enabled life on earth to leave the seas and spread over the land.

It's evident that Tatooine has oxygen. Otherwise, Luke and his family would not be able to breathe there unaided. The presence of the oxygen could also support a protective ozone layer on Tatooine. However, since there appears to be a serious lack of plant life on Tatooine, this begs the question: Where is the oxygen coming from?

Ruling out the possibility of vast colonies of oxygen-producing bacteria existing somewhere on the planet, there has to be another option. It's possible that the oxygen was produced in the past when Tatooine had a climate that could support a great deal of plant life.

It's been calculated that if photosynthesis suddenly stopped and no more oxygen were released on Earth, it would take all living things more than thirty-five thousand years to deplete the oxygen through respiration. So we could potentially be seeing Tatooine within the thirty-five thousand year period after the loss of its main oxygen producers.

So life could have evolved on Tatooine in the past, before it became a desert planet. If it was always a desert, it's unlikely that life would have become as complex as Jawas. However, in regard to potential life on desert planets in general, Azua-Bustos says:

> *"We do know that there are a number of planets in binary systems. We do know also that life is able to persist in extremely dry places, like the hyperarid core of the Atacama Desert. So if some of these planets do have an atmosphere with water in it, they may as well be inhabited."*

HOW LONG AGO AND HOW FAR AWAY?

"A long time ago, in a galaxy far, far away . . ."

The classic entrance to the movie that gives the impression that this could be a tale about a world that once was, and maybe even still exists somewhere in the far reaches of our known universe. We're fortunate enough to be alive at a time when people have worked out that we live in a huge universe that's been around for an extremely long time.

As far as we know, the universe contains everything that has or will ever happen, including time. If we were to try and place Star Wars within a time or place, then a good place to start would be to work out the extremes of the known universe.

How Long Ago?

The current theory that describes the origin and evolution of the universe is called The Big Bang. In this model, time and the universe began 13.8 billion years ago. Since then, both have been growing continuously larger.

In the beginning, it was extremely hot. Everything in the universe was squashed into an area (called a "singularity") many times smaller than the period that ends this sentence. So the energy density was immense. The resulting temperature was so high that even the particles that make atoms couldn't yet form. However, by the first second, the universe had expanded and cooled sufficiently to allow protons and neutrons to form, followed closely by electrons. These particles are the basis of the atoms that make up (most of) everything around us.

For the first twenty minutes of the universe's life, there were only simple nuclei. These nuclei contained just one or two protons, which we recognize as hydrogen and helium, respectively. About three-quarters of the universe's mass was hydrogen nuclei, and the remaining quarter was mostly helium nuclei. It wouldn't be until about 377,000 years in that the universe would cool enough for electrons to start bonding to the nuclei to make atoms of hydrogen, helium, and traces of lithium (which has three protons in its nucleus).

Humans are made of only about 9.5% hydrogen by mass. We don't contain helium. The rest of us is made up of 65% oxygen atoms, 18.5% carbon, and varying amounts of more than fifty other elements. All of these heavier elements weren't present in the first few hundred thousand years of the universe. They needed to be made by a process other than The Big Bang. They were made in stars.

Starstuff

Stars produce heat and light through the fusing of atomic nuclei together to create other more heavy nuclei. Over enough time, a star can convert its hydrogen into helium, carbon, oxygen, and iron as well as other elements in between. Depending on the elements that a star begins with, they are classified into three main populations.

Population III are considered as the oldest. These hypothetical stars mainly convert hydrogen, helium, and lithium into heavier elements. At the end of their life they explode, seeding space with their newly synthesized elements, which are eventually used for making other, younger stars.

Population II stars are the oldest observable stars (some are thirteen billion years old) and are thought to have created most of the other elements in the universe. They amount to about 39% of the stars in our Galaxy. These stars formed early in the universe's history and are considered metal-poor. In astrophysics, metals are any elements with three or more protons, i.e., not hydrogen or helium.

Population I stars are young, metal-rich stars that were present as early as 10 billion years ago. The Sun is an intermediate Population I star, found further out in the Galaxy; however, the youngest, extreme Population I stars exist closer to the center of the Galaxy.

If we use our Sun as a successful example of a life-bearing solar system, then we're looking for a Population I star. The oldest are about 5.5 billion years older than our Sun. The planets in our solar system are also 4.5 billion years old, which is in line with the Sun. Planets have a similar age to their parent star as they form around the same time and from the same cloud of dust and gas that the star is formed.

Out of the many bodies orbiting in our solar system, Earth is the only planet that we are certain has harbored life. Once Earth formed, it took as long as one billion years for life to emerge. So we can imagine that life could have made a similar appearance on one or more of the earliest Population I stars. So life could have appeared as much as nine billion years ago, according to this reasoning.

The appearance of life isn't the end, though. From microorganisms to mammals, there can be no Star Wars unless there is a species that can communicate, cooperate, and create technology to get them to the stars. On Earth, it took the majority of life's history before species evolved into life-forms capable of space travel. That's at least 3.5 billion years! Currently there is no way to tell whether this would be the norm on other planets, considering the amount of cosmic and terrestrial coincidences that have occurred to bring us to our current ecological state.

Life has seen many catastrophes from space and within. The Great Oxygenation event occurred as early as 2.3 billion years ago when oxygen-producing bacteria raised the global abundance of free oxygen to levels that were poisonous to existing life-forms on Earth. We are now dependent on that oxygen to live. When an asteroid crashed into the Gulf of Mexico around sixty-six million years ago, it prompted the extinction of the dinosaurs, making way for mammals to flourish, and ultimately led to the emergence of modern humans as much as two hundred thousand years ago.

Humans first left the planet in the 1960s, but we still don't know how long it will take humans to be able to travel freely amongst the stars, or whether it will ever be possible.

So let's assume we develop the technology to visit other star systems within the next thousand years. We can say it took roughly 3.5 billion years for life to become star-worthy after a star appeared that had all the necessary stuff for life. Considering this was as early as ten billion years ago, this means that potentially one or more galactic life-forms could have developed star travel as long as 6.5 billion years ago.

Finally, we can't have Star Wars with only one side to the battle. There either needs to be life-forms capable of developing on a multitude of planets in the Galaxy, or once life was able to traverse the stars, it had to have enough time to spread to other suitable star systems and gradually evolve into the many different forms we see in the Star Wars universe.

You might ask: How spread out is everything anyway?

How Far?

Space is big. In fact it's so big that to be able to better convey the distances we need to resort to comparing distances to the speed of light. We measure the distances by how long it would take light to get there.

Light is the fastest thing we know of; in space, it can travel 299,792 kilometers in one second. This means it can get to the moon in about 1.3 seconds. We would say that the moon is 1.3 light-seconds away.

As mentioned in a previous chapter, the distance from the Earth to our nearest star, the Sun, is known as an Astronomical Unit (AU). 1 AU is a distance of 149.6 million kilometers or 499 light seconds (8.3 light minutes). The distance to our furthest planet, Neptune, is 30 AU, and the edge of the Sun's system, the heliopause, is more than 120 AU away. That's a distance of 18 billion kilometers, or 1,000 light minutes (16.7 light hours).

Now, 16.7 hours for light to get to the heliopause might seem like a long time, but consider the distance to the nearest star other than our Sun. It's called Proxima Centauri and it's 4.2 light-years away. At this distance, kilometers and AU stop being useful, so astronomers may resort to using the parsec. A parsec is used to measure distances between stars and is equivalent to a distance of 3.26 light-years. As such, the Milky Way is 33,726 parsecs across, meaning it takes 110,000 light-years for light to cross our Galaxy.

The Milky Way is a spiral galaxy, which is how the Star Wars galaxy is also portrayed. More than two-thirds of the known galaxies in our universe are spiral galaxies. The nearest to the Milky Way is called the Andromeda Galaxy, which is a staggering 2.5 million light-years away. This is almost twenty-three times the width of our Galaxy. So how many galaxies are there that could be suitable locations for the Star Wars galaxy? It's been estimated that there are well over one hundred billion galaxies out there.

In 2004, NASA unveiled the deepest image of the observable universe ever made by mankind. Called the Hubble Ultra-Deep Field (HUDF), it focused on a dark area of the sky, roughly one-tenth the diameter of the moon, and took a one-million-second long exposure. It revealed ten thousand galaxies in that area alone.

When looking into space, we are limited to what we can observe. This is called the observable universe and is limited by how far light has been able to travel to us since the beginning of the universe. Even so, since the early twentieth century we have known that galaxies are moving away from each other in what is referred to as the Hubble Expansion, named after American astronomer Edwin Hubble.

It is now well-documented that the universe is expanding, so much so that the furthest galaxies are much further away than they appear to be. In the time it has taken the light to reach us from them, they have moved further away. The current estimate for the proper size of the observable universe is about ninety-three billion light-years across.

Just like finding stars made of the necessary elements for life, the same applies to galaxies. Some galaxies are young, while some were forming near the beginning of the universe. These early galaxies weren't spirals, though; they were elliptical or irregular galaxies. The earliest example is GN-z11, thought to be located 13.4 billion light-years away.

When looking at the most distant spiral galaxies so far observed, such as BX442, the light has taken 10.7 billion years to get to us. However, the expansion of the universe gives them a potential proper distance of more than thirty billion light-years.

So how long ago and how far away could the Star Wars galaxy have existed? The answer: approximately six-and-a-half billion years ago, and more than 13.4 billion light-years away.

HOW COULD WE LIVE ON A GAS GIANT LIKE BESPIN?

So your hyperdrive has failed and you're left with sub-light engines. You need a safe port within a reasonable distance to lay low for a while. Your best bet? A mining colony on the giant gas planet Bespin, run by your old buddy Lando.

On arrival you're intercepted by a stern team of guards in two cloud cars ready to accompany you to the floating metropolis that is Cloud City.

Flying over this 16-kilometer-in-diameter city, you arrive and land on Platform 327, where you exit your ship to take in a deep breath of the Bespin atmosphere. Fortunately for you, Han Solo, Chewbacca, and Princess Leia, the Bespin atmosphere is breathable at this altitude.

Now in the film, Han, Chewie, and Leia don't appear to put on an extra layer of clothes when they disembark the *Millennium Falcon*. So, although it's a bit breezy, the climate on the city doesn't appear to be too cold.

Considering that this all takes place amongst the clouds of a gas giant, the climate seems a bit too comfortable to be true. What would it really be like to inhabit a gas giant like Bespin?

Bespin

Bespin is a planet similar to Jupiter and Saturn in our solar system. With a diameter of about 118,000 kilometers it is almost the same size as Saturn.

It's generally thought that gas giants are so big due to their distance from the Sun when they formed. As they were forming, these soon-to-be behemoths collected up matter that lay within their orbits. Being further out means that there is a larger orbit, and thus more material from which to form the planet.

After forming, planets can also migrate towards or away from their sun over millions of years, which may have been the case with Bespin.

Closer to a star, where temperatures are high, light, volatile molecules escape as gas. Further away, these volatile molecules freeze and can be captured by the forming planets. Gas giants are composed of mainly hydrogen and helium, which are the lightest elements.

On the other hand, the inner planets of our solar system are made of mainly heavier, less-volatile molecules and elements that can exist as liquids or solids at higher temperatures. It's these heavier materials that are commonly mined on planet Earth, and in the future, potentially on asteroids and our moon.

On Cloud City they mine the lucrative Tibanna, a rare gas that exists within the upper atmosphere of Bespin. Having a valuable resource to trade makes Cloud City a wealthy place to operate from, and means there is a fair bit of traffic to and from it. This allows them access to regular visitors from different star systems. These visitors are sources from which residents can obtain money as well as needed supplies.

Bespin also has several moons, which is a trait common to gas giants. In our own solar system, both gas giants have more than fifty confirmed moons; some bigger than Mercury. Their features include thick atmospheres, active volcanoes, deep oceans, and even cryovolcanoes (ice-spewing volcanoes).

If the moons around Bespin are anything like the moons around Saturn, then they could be directly exploited for useful materials such as water and organic molecules. These could be harvested and brought to Bespin by more local operations, thus reducing costs to Cloud City.

The Floating City

As previously mentioned, Cloud City is a mining colony and host to millions of tourists. Like many things in Star Wars, it's enormous! It has 392 levels, the uppermost of which are used as a luxury resort; the lower of which are used to house staff and process the Tibanna gas.

There are many problems to overcome when inhabiting such great altitudes. There's the air pressure, the temperature, and the need for breathable air, in addition to the obvious peril of making sure your city doesn't fall out of the sky.

Cloud City isn't the first floating city in science fiction. Many years earlier Jonathan Swift introduced us to the floating island of Laputa in his book, *Gulliver's Travels*. Laputa was kept afloat by means of a huge Lodestone (a natural magnet), whereas Cloud City is said to use repulsorlift engines and tractor beam generators to stay aloft.

If we want to defy gravity by hovering, we're currently limited to technologies such as lighter-than-air crafts, vehicles with ducted fans (like toy drones), rotary wings, or jet propulsion. These technologies only work up to a maximum altitude or ceiling, above which there isn't enough air pressure for them to function properly.

The operational ceiling of repulsorlift technology isn't based on air pressure, though. It's meant to work by pushing against a planet's gravity. So it's operational ceiling would be the altitude at which the gravity is too weak for the repulsorlift engines to support the city's weight.

This antigravity technology is ubiquitous in the Star Wars universe, but nothing like it exists in the real world. In recent years, some organizations have given it some serious thought, though. This includes BAE Systems, which supported an antigravity project called Greenglow. As of right now, though—until a breakthrough in antigravity technology—it looks as if the idea of a floating city just isn't feasible.

Atmosphere

Jupiter and Saturn are different sizes and distances from the Sun.

Jupiter's average cloud-top temperature is -108°C compared to Saturn's -180°C. Both planets are composed of more than 98% hydrogen and helium, although Saturn has a slightly bigger ratio between its hydrogen and helium. The remaining portions of both planets' atmospheres are mostly methane with traces of ammonia and water vapor, as well as other gases.

If you descend through the atmosphere, the pressure and density increase. Below a two-hundred-kilometer outer cloud layer, you'd hit one thousand kilometers of gaseous hydrogen above a sea of liquid hydrogen. Any deeper and the pressure would crush your ship. This explains why Cloud City is located in the upper atmosphere, fifty-nine thousand kilometers above Bespin's core. However, the temperature and pressures of such an environment can also pose problems.

On Earth's surface we experience one bar of pressure. However, it decreases as we go up higher. This reduces the amount of oxygen available to us in each breath. The temperature also decreases rapidly the higher you ascend, diminishing with a lapse rate of almost 10°C per kilometer. This is why mountaintops are frequently covered in snow.

In mountaineering there's an area called the Death Zone, where there's insufficient oxygen to breathe. A person would die in a few minutes without an oxygen bottle and adequate thermal clothing. On Bespin, there is conversely a region called the Life Zone, where the opposite is the case. It's described as a region in the Bespin atmosphere that is breathable to humans. This is where Cloud City is situated.

The Life Zone

On a gas giant like Bespin, to get a layer of atmosphere that contains enough oxygen is a major problem. On Earth, the oxygen supplies come from life-forms that use photosynthesis to obtain carbon from carbon dioxide (CO_2), leaving the oxygen to enter the atmosphere. By volume, oxygen is only about 21% of the air we breathe. Nitrogen makes up the majority at 78%.

Bespin would need to have similar photosynthesizing life-forms such as algae to oxygenate its Life Zone, unless it has another way of obtaining oxygen. However, the algae would need access to water, which is only in the clouds. They would also need a source of CO_2 which, if present on the planet, would just sink to the lower atmosphere due to its weight.

Astrophysicist Mathieu Hirtzig, who now works at Fondation La main à la pâte in Paris, has spent a significant amount of time studying Saturn's moon Titan. He proposed some possible alternatives for oxygen production based on stellar flux (energy from the local star) and technology.

"Photolysis (the breaking of molecules by UV rays) is an option: the water molecule could be broken down by the stellar flux, to keep only Oxygen," Says Hirtzig. However, because Bespin is massive enough to hold onto any freed up hydrogen, the oxygen would just reform into water or even more reactive molecules.

The technological option is to use electrolysis where the water is split using electricity.

"The Cloud City could well be producing its own Life Zone, by keeping the hydrogen and using it as fuel. [This could] liberate oxygen in its surroundings for habitability."

Regarding the process of filling the Life Zone with oxygen, Hirtzig stated that "it would be more efficient to keep it in tanks and leave it inside the city only."

The Life Zone would need a whole lot of oxygen to fill it to the 21% oxygen level that we breathe on Earth. If the Life Zone were ten kilometers deep, centered on an altitude of fifty-nine kilometers, it would contain seventeen times more volume than the entire Earth's atmosphere up to and including the stratosphere.

Handling the Pressure

Saturn has different cloud layers at different heights. The highest layers have ammonia clouds at pressures similar to Earth's surface. Then ice clouds are lower in the atmosphere, at pressures between 2.5 and 9.5 bars and no warmer than -3°C. Then if we are brave enough to descend to pressures of 10 to 20 bars, we can find water clouds at temperatures from -3°C to 57°C.

So the upper atmosphere of Bespin would be extremely cold. In the regions where the pressure matches Earth's atmosphere, the temperature would be way below freezing. To get to a temperature that wouldn't freeze your bits off, the pressure would be at least ten times more than we experience on Earth.

Our bodies can take high pressures, as evidenced by underwater divers, but at these high pressures the oxygen and nitrogen in the air become toxic to us. Therefore, any Life Zone on Bespin would have to have oxygen existing at a partial pressure that's suitable to breathing. This is roughly between 0.16 and 1.6 bars. However, again, any oxygen would just sink in the atmosphere instead of being held within a layer.

In conclusion, it seems highly unlikely that a gas giant like Bespin could support life as portrayed on screen. Matthew Hirtzig suggests that heating up a smaller and colder ice giant may have been more plausible.

PART III
ALIENS

WOULD AN EXOGORTH REALLY EVOLVE ON AN ASTEROID?

The question of alien life has given us some of the best movie taglines of all time: "In space, no one can hear you scream." "We are not alone." And, of course, "A long time ago, in a galaxy far, far away." Science fiction writers and directors, like George Lucas and J. J. Abrams, have thought long and hard about how to portray creatures from other planets.

The Star Wars universe has gifted us a cornucopia of alien creatures to examine. The Tusken Raiders, aka Sand People, may walk like humans, but they are vicious monsters that stalk the sands of Tatooine. The Hutts are a sentient species of large gastropods. Often seen as crime lords, the Hutts come with short arms, huge eyes, and wide, cavernous mouths—all the better to eat you with. The Gungans are an intelligent race of amphibious humanoid aliens native to the waters of Naboo.

Each alien is fit for the environment in which they dwell.

Darwin Invented the Alien

It was Charles Darwin who first dreamt up the modern alien we see in Star Wars. Darwin's theory of evolution gave fiction a science of imagining the ways in which life might arise on other worlds, as well as ours. Before Darwin, extraterrestrials were not genuine alien beings. They were merely dudes and dudettes like us, living on other planets.

But Darwin changed all of that. From Darwin on, the notion of life beyond our home planet became truly alien, and associated with the physical and mental characteristics of the true extraterrestrial. Through science fiction, especially Star Wars, the idea of the alien became deeply embedded in the public imagination.

Archetypal aliens soon developed: alien as highly evolved killer, alien as ocean-planet, and alien as wise and benevolent mentor.

For the alien as highly evolved killer, witness the Martians in *The War of the Worlds* (2005), or the predatory xenomorph in Ridley Scott's *Alien* (1979). The Tusken Raiders might fit into this category.

The alien as ocean-planet was famously portrayed Stanisław Lem's novel *Solaris* (1961), which was later adapted into film (1972, 2002). The swirling sentient sea in *Solaris* is a single organism with a vast, yet strange, intelligence that humans strive to understand. Star Wars has its own sentient planet in Zonama Sekot, which is capable of independently traveling through space.

And the alien as wise and benevolent mentor? This conjures up memories of the civilized aliens of superior intelligence in films such as *Close Encounters of the Third Kind* (1977). And, of course, Yoda, a mentor who possesses almost-infinite wisdom.

Carbon Versus Silicon

But what of the exogorth, the giant space slug that resided in the hollows of asteroids?

The Star Wars legend has it that the exogorth was a silicon-based species, a member of which was seen living in the Hoth asteroid belt. Let's pause and consider that for a moment. Sci-fi is often seduced by silicon. The xenomorphs from the Alien franchise are apparently silicon-based. As are the Tholians of *Star Trek* and the creature in the *X-Files* episode, "Firewalker." Silicon is chemically similar to carbon, the chemical element upon which life on Earth is based. The thinking goes something like this: If carbon and silicon are so similar, and carbon begets life, why shouldn't silicon also do so? Those who resist the temptation to be seduced by the silicon argument are often called "carbon chauvinists."

However, perhaps there's a good reason to admire carbon.

Carbon is cosmic. It forms the basis of life on Earth, and is also to be found out in deep space. Carbon is the backbone of biology because of its very nature. It easily bonds with life's other main elements, like hydrogen, oxygen, and nitrogen. Carbon is also light and small, making it an ideal element for creating the longer and more complex chemicals of life, such as proteins and DNA. Carbon also makes one of the *softest* known substances in graphite, and one of the *hardest* known substances in diamond.

All in all, carbon is known to form ten million different chemicals, which is the majority of all chemicals on Earth.

Yet, the silicon cheerleaders ask us to imagine a world on which silicon is abundant. In such an environment, would silicon not replace carbon as the chemical fit for life? Indeed, there *is* such a planet. It's called Earth. Silicon is more abundant than carbon on the Earth's surface, and yet terrestrial life is almost exclusively carbon-based.

Silicon chemistry notwithstanding, consider the exogorth further. When we look hard into the history of sci-fi, we find such creatures have a noble pedigree.

The famous medieval German astronomer Johannes Kepler also imagined creatures on a hostile world. Kepler was the guy who first nailed the planetary laws that govern our solar system, and after whom NASA's *Kepler* planet-hunter telescope was named.

Kepler had written one of the very first works of science fiction in *Somnium*, his 1634 book about a journey to the Moon. He was one of the first writers to imagine alien life. The extraterrestrials that stalk Kepler's Moon are not humans. They are creatures fit to survive their alien haunt. Two centuries before Darwin, Kepler had intuitively suggested a bond between life-forms and habitat.

And what Moon creatures did Kepler imagine? Serpents. Like the exogorth on the asteroid near Hoth, Kepler's serpents face a harsh existence, with extremes of light and dark, hot and cold. The serpents bask for a fleeting instant in the rising or setting sun, then creep into the impenetrable darkness. However, Kepler had assumed there would be some kind of atmosphere on the Moon, and that looks unlikely on our famous Star Wars asteroid.

Exogorth Habitat

The exogorth would have to evolve to suit its situation. How would it survive?

According to Star Wars legend, the exogorths reach full maturity at ten meters and reproduce by splitting into two smaller, separate bodies. If an exogorth was unable to separate, its growth would continue unimpeded, potentially reaching nine hundred meters in length. Exogorths of this size are said to have swallowed spaceships whole, and house entire ecosystems.

Apparently, exogorths often lurk in asteroid fields. They burrow fully into an asteroid until completely hidden. From the murky darkness, they feed off stellar energy emissions, mineral-rich deposits within the asteroid, and floating space debris. Largely dormant, the opportunistic feeder could lunge upon passing ships, though doing so would apparently use up its energy, exhausting it.

On Earth, we also have creatures that subsist in subterranean settings, surviving on a diet of rock and water. Some asteroids are thought to have a veneer of permafrost, as well as having experienced periods in the past where their interiors had become molten. However, it's doubtful that the inside of an asteroid would have remained melted for long enough for life to emerge.

Perhaps there's much in Han Solo's remark that anyone would be crazy to follow him into an asteroid field. It's possible that such folklore recognizes that past victims had been swallowed whole by lurking exogorths. It's even possible other ships that had previously attempted to pass through the rocky field were smashed into the smithereens that made up the space debris in the exogorth diet, much like sky whales, feeding on cosmic plankton.

But basking on cosmic plankton won't give the exogorth incredibly massive teeth.

Those gnashers would surely come from the exogorth's "daily" diet of munching on other silicon-based creatures, such as mynocks. These were a species of silicon-based bat-like parasites, which chewed on the power cables of starships and, like the exogorths, were capable of surviving in the vacuum of space.

Water Bears!

On Earth, we have our own creature that can survive the vacuum of space. They're called tardigrades, or water bears.

Tardigrades are essentially indestructible. They don't look particularly impressive, with sofa-like bodies and four pairs of stubby, poorly articulated legs. They're certainly not of exogorth dimensions, or even as massive as a mynock. Tardigrades range in size from 0.012 to 0.020 inches, though the largest species may reach a heady 0.047 inches. Scientists have found tardigrades on top of Mount Everest, in hot springs, under layers of solid ice, and in ocean sediments.

They're able to survive the most extreme environments that would kill almost every other animal. They've survived temperatures as hot as 151°C, and as cold as -272°C. They can go without water for ten years. They can stand one thousand times more radiation than other animals. And they've even been known to survive the vacuum of space. For ten days in 2007, tardigrades were taken aboard the FOTON-M3 mission into low Earth orbit, where they were exposed to the hard vacuum of outer space. Did that faze them? Not in the slightest. On their return to Earth, and after rehydration, most of the tardigrades recovered within thirty minutes.

So it may be wise not to dismiss the exogorth.

Writers and directors, as well as scientists, are pushing the boundaries of how life might arise and survive in space. In a universe where Zonama Sekot can travel freely through space, what price a toothed gastropod that survives by feasting on rock and metal?

HOW COULD REY SURVIVE ON THE DESERT PLANET OF JAKKU?

Jakku.

The metal hatch of a junked space vessel is snatched open to show the blue-lit, mummified face of an alien scavenger, decked out in bug-eyed goggles, keffiyeh-covered face, and gloves.

As the camera pulls back, we get a full-bodied look at the alien. It is well-equipped. Its expedition backpack has a full-length combat staff strapped on. The creature seems tooled-up for every task. We are in an upside-down, canted corridor. The scavenger finds a precious piece, drops it in a satchel. The satchel is swung back and the scavenger clambers down a cable, flanked by estranging walls of machinery.

Alone and elfin in this gargantuan space of a sideways shipwreck, the scavenger descends down the two-hundred foot cable, landing hard on rusted metal, and heads through the ferric dust toward a distant slit of sunlight.

Emerging from the machine darkness, the scavenger pulls off the bug-eyes to reveal the flushed and determined face of a beautiful, young, female humanoid. She holds a canteen to her lips, and shakes the last drops of water into her mouth.

Rey.

Cut wide, to reveal a miniscule Rey set against the epic engine of a crashed star destroyer. Rey sets off down a sand dune and onto the salt flats below. Her speeder is dwarfed against a graveyard of wrecked starships as she sputters to a remote outpost of civilization.

Jakku

Jakku is an isolated desert world located in the Western reaches of the Star Wars galaxy. It is a hostile planet in which Rey managed to survive. Deserts are habitats of extreme climate, where heat exhaustion is a real threat, and where rainfall is often less than ten inches in a year. A harsh environment, a desert planet like Jakku is only survivable by the hardiest of humans. And it's clear that Rey qualified.

So, how does this planet of Jakku—this definitive desert world of dry, sandy landscape, searing skies, and highly elevated temperatures—compare with what we know today?

On Earth, such a landscape exists in the form of the Sahara in North Africa, part of a desert band which stretches through the Middle East and into south and central Asia. The driest place on Earth, however, is the Atacama Desert in South America. Some parts of the Atacama have never seen rain. The high mountains that divide the desert from the Amazon jungle create an obstacle for rain clouds. All the rain drops on the mountains, placing the Atacama in their rain shadow.

On Earth, too, there are sand dunes. Our home planet boasts coastal dunes and land-locked desert dunes, laid out in pretty patterns. There are even sand dunes on Mars. Both planets have the kind of towering sand dunes that Rey rode down to the salt flats.

Sand forms from the erosion of rocks. Through wind and time, a dune can accrete to a mountain as high as a thousand feet. Skyscrapers in sand. Dunes are far from desolate stacks of lifeless silica. They are dynamic natural structures. Sands shift, grow, and travel. Dunes crawl with the biota of their habitat.

Some Jakku dunes may even have sung.

Dunes sing on Earth, too. In Morocco and Chile and Nevada, booming sand dunes sing in deep hums that can resonate for more than a few minutes. The melody stems from sand-slips, the eerie sounds reverberating when the dry sand slides like an avalanche. Rey may even have made the sand sing by sliding down the slip-face of a dune. As the sand began to vibrate, she'd have heard the dune singing.

For Rey, sandboarding on Jakku would have been the same as that on Earth or Mars. Sandboarding is similar to snowboarding; same board, dissimilar surface. Sand is lighter than snow, so boarders get a swifter ride than they would on snow. You can sandboard with your snowboard, or you can buy a sandboard, which has a far glassier base. Rey made do with sheet metal. It was a way of getting down the sand dune slickly.

A sand dune can swallow you whole. So, for Rey, sandboarding would have been no mere sport.

Loose sand can be deadly. And deteriorating dunes are unstable and can collapse unexpectedly. Most of the time, it's safe to make your way across the miles of endless dunes, but on Jakku, it is always advisable to be well-equipped and to notify someone of your plans, especially if traveling through the Sinking Fields in the Goazon Badlands. The Fields were located north of Rey's AT-AT home, and they were the site of the crash landing by Poe Dameron after escaping the First Order's flagship, *Finalizer*. Soon after impact, the Fields swallowed Poe's TIE fighter whole, to the horror of Finn.

Surviving the Desert

Rey had a limited chance of survival in the desert.

Her main challenges were the hostile heat, and the few natural resources available. Shelter and water were scarce. Vegetation could be hard to find, as many creatures hid from the heat in the daytime. Vegetated areas (known as wadis on Earth), grassed plains, and higher ground were more promising places for survival. Wadis may hold water from what rain may have fallen, and higher ground usually meant colder temperatures, as well as good vantage points and visibility. However, the greatest danger came from dehydration and heat exhaustion. Rey kept water on her person at all times.

There was also the prospect of predators, such as the ripper-raptor and the Arconan night terror.

The ripper-raptor was a little-documented species of flying reptile, with leathery wings and keen eyesight, perfect for the vast plains of

Jakku. Ripper-raptors were flesh-eaters that rode the thermals in the deep of the desert in the Kelvin Ravine, trying to pick off inhabitants of the settlement of Tuanul. The Arconan night terror, also known as the Nightwatcher worm, was a sand-boring specimen with large, red eyes, and a body that could reach twenty meters in length, though some were reputed to be even longer. On Jakku, scavenger lore told tales of Nightwatchers skulking motionless under the sand, only to spring and snap at their prey at the slightest vibration on the surface. So Rey made sure she was armed and ready.

Making Ends Meet

Rey would have been drawn to the Steelpecker.

Spotted in an early scene in *The Force Awakens* as Rey's speeder first sputters across the sands, the Steelpecker was a non-sentient carrion bird with beak and talons tipped with iron. They fed mainly on metal, which meant their digestion systems stored vanadium, osmiridium, and corundum in their gizzards. All this made the Steelpecker a very useful commodity to scavengers like Rey, who collected their carcasses and guano.

But Rey had more civilized ways of making ends meet.

Jakku's place in unpopulated space meant that the planet could be used as a jumping-off point for warships heading west, into the Unknown Regions. At peak points, such as the Galactic Civil War, this meant increased traffic and, if scavengers were lucky, more wrecked starships. Even if the wreckage was to be found away from settlements such as Tuanul, Cratertown, or Reestkii, Luggabeasts could be used to carry supplies across the desert to Niima Outpost, a trading post and the only major settlement on the planet.

Disabled and discarded engine parts can easily be fashioned into something creative, as witnessed by the way Anakin managed to build an entire podracer from junked parts. So, Rey would scrub her day's salvage clean, and present it by means of trade to her blobfish boss, Unkar Plutt. After scanning the salvage in valuation, the wreckage would be traded for sealed packets of dried green meat, or powdered bread.

Desert Planet Study

A couple of summary questions occur. First, is it really possible to have a whole planet consisting of sand? And second, isn't the question of survival on such a planet slightly academic anyhow?

A 2011 study found that not only are life-sustaining desert planets possible, but that they may even be more common than Earth-like planets. Models made by the scientists in the study found that desert planets had a much larger Goldilocks Zone than watery planets. The study also suggested that Venus may once have been a habitable desert planet, as recently as a billion years back, and that the Earth may become a desert planet within a billion years due to the Sun's increasing luminosity.

Perhaps, we'd best pay Rey's survival skills more attention.

DOES THE FORCE AWAKENS PASS THE TEST FOR AN ALIEN HEAD COUNT?

The *Millennium Falcon* drops out of lightspeed, and alights upon the sumptuously lush blue-green planet of Takodana.

Situated on the trade routes between the Inner and Outer Rim, the planet was a popular departure point, perfectly placed for those heading out to the galactic periphery—a last taste of civilization. Little wonder the castle's bizarre wide-walled saloon is a shop of little horrors, full of rough alien travelers who gamble, drink, negotiate, and scheme. An "incredible collection of extraterrestrial alcoholics and bug-eyed martini drinkers lined up at the bar," American film critic Roger Ebert said of a similar collection of aliens in another cantina, on another planet, exhibiting "characteristics that were universally human, I found myself feeling a combination of admiration and delight."

Back in *The Force Awakens*, our roving lens ends up on a tiny, one-thousand-year-old female alien who—wearing her large, adjustable goggles—looks like a cross between Ichabod Crane and Z, the worker ant. This is Maz Kanata, bug-eyed pirate-queen proprietor of the bizarre saloon.

That saloon in Maz's castle on Takodana is a reminder of just how replete with alien life the Star Wars galaxy is—up to twenty million different sentient species, if you listen to Legends, whereas canon is characteristically quiet on putting a number to the galactic population.

Is Star Wars overly optimistic about the diversity of life in the cosmos? Do we have a way of estimating how much life there may be in our own

Galaxy? Does *The Force Awakens* pass the best "alien head count test" scientists have to offer?

The Drake Equation

Scientists know that "alien head count test" as the Drake Equation.

The equation is named after American astronomer, Frank Drake, who in 1960 carried out Earth's first search for alien life when he trained an eighty-five-foot radio telescope on two sun-like stars in the solar neighborhood. The equation was written in 1961, and was prepared for a meeting at the National Radio Astronomy Observatory in Green Bank, West Virginia. Frank was joined at the meeting by some serious heavyweight scientists, including legendary American astronomer and prolific science author Carl Sagan, Nobel-prize winning chemist Melvin Calvin, eminent radio-astronomer Otto Struve, visionary neuroscientist John C. Lilly, and Manhattan Project physicist Philip Morrison. The attendees, ten in all, dubbed themselves "The Order of the Dolphin," after Lilly's work on dolphin communication.

For those hardy of mathematical heart, the equation looks like this:

$$N = R_* \bullet f_p \bullet n_e \bullet f_l \bullet f_i \bullet f_c \bullet L$$

Each variable symbolizes the following:

N = the number of communicative civilizations in our Galaxy

R_* = the average rate of star formation in our Galaxy

f_p = the fraction of those stars that have planets

n_e = the average number of planets that can potentially support life

f_l = the fraction of planets that actually develop life

f_i = the fraction of planets with life on which intelligence arises (civilizations)

f_c = the fraction of civilizations that develop detectable signs of communication

L = the length of time such civilizations send communicative signals into the Galaxy

The idea is this: to find out how many alien civilizations (N) might be communicating between the stars, out in Galactic space, you get an estimated number for each of the factors, and multiply them together. Simple!

The Alien Head Count

The equation cleaves into fact and fiction.

Roughly speaking, the first five factors on the right hand side of the equals sign of the Drake equation (R_*, f_p, n_e, f_l, and f_i) are mostly questions of quantity and science. The last two factors (f_c and L) are more concerned with the qualities of human politics and anthropology.

Star Wars fiction has more to say about the last two factors of the Drake equation than science does. The rise and fall of galactic civilizations is the business of franchises, and not yet the concern of astrophysics or evolutionary biology.

Better than ever before, experts are in a good position to put numbers to the first five factors. In 1977, science knew of no planets in orbit about other stars. It wasn't even very clued up on the gas giant worlds in our *own* outer solar system. Even in 1999, the year the first of the Star Wars prequels hit cinema screens, scientists were aware of only a limited number of gas giants around other stars. But today, thanks to NASA's *Kepler* space probe, experts estimate there may be as many as four and a half billion Earth-like planets in orbit around just the red dwarf stars in our Galaxy.

Yet, there's still a problem with those five science factors in Drake.

Two scientists, both using the same data sets and both unaware of logical error, can reach wildly differing conclusions. Those conclusions can even be as stark as this. One would believe the data set showed that Star Wars has got it right. Our Galaxy is indeed also full of communicating civilizations. We just haven't met up with them yet. Or maybe they're on the other side of our Galaxy. Whereas the second scientist concludes we're alone in the universe!

Aliens in Space and Time

So some of the Drake dynamics can be reckoned at, while other factors are far more like fictional guesswork.

But when scientists *do* come to a consensus, something fascinating emerges, something very interesting for our consideration of the alien head count in the Star Wars galaxy. And it's this: when all the "best guess" data has been factored into the Drake equation, the number of communicating civilizations in a galaxy (*N*) turns out almost equal to their average lifetime in years.

For example, say a typical alien society stays on (cosmic) line for a mere one hundred years, then there will only be roughly one hundred of them in the galaxy. In which case, your chance of meeting up with Balosars, Mynocks, and bug-eyed monsters is pretty slim. On the other hand, if extraterrestrial civilizations last a good ten thousand years, the stats suggest there will be roughly ten thousand such societies, and alien contact is a much better bet.

So our focus should shift towards time, and away from space.

That vast illimitable ocean of a Star Wars galaxy, in which up to twenty million species may swim, simply means that the average lifetime of an alien civilization in Star Wars is twenty million years. That doesn't seem excessive, in a universe clocking in at 13,000 million years old.

If there's rank optimism in the portrayal of life in the Star Wars galaxy, it's twofold. One, the Star Wars spin on those science factors in Drake (R_*, f_p, n_e, f_l, and f_i) is admittedly on the high end of allowable values. In other words, the Star Wars galaxy has a high number of life-bearing, planet-bearing stars, which go on to evolve intelligences capable of communicating with other star systems. Two, Star Wars has faith in extraterrestrial intelligence. There's a belief that "intelligence" can overcome the politics of greed, in fouling the planetary nest, or the politics of war, in blowing themselves up.

Another Take on Drake

It's not an original thought to conclude there are problems with the Drake equation.

From the get-go, Frank Drake only ever suggested its seven parameters were a way of better framing the Search for Extraterrestrial Intelligence. However, a new study in 2016 from the universities of Rochester and Washington has attempted to make amends for the Drake approximations. The study has also tried to put an exact number on the likelihood that humans are alone in the universe.

The study found that humans would be alone in the universe *only* if the odds of alien intelligence developing elsewhere are less than about one in ten billion trillion, or one in 10,000,000,000,000,000,000,000. In short, it's extremely likely that there is, or has been, intelligent life elsewhere in space.

Now, admittedly, this new study doesn't help with the alien head count in Star Wars, nor does it answer the perennial question as to whether the "truth is out there" with regard to the existence of aliens. But it *does* come to a conclusion in its own right: aliens have existed.

And those aliens could have existed a long time ago, in a galaxy, far, far away. . . .

HOW COME MANY PLANETS IN STAR WARS HAVE BREATHABLE ATMOSPHERES?

Lost in space.

A group of human colonists left their home planet of Grizmallt and headed out to the stars. Their journey foundered in the Mid Rim, close to the border of the Outer Rim Territories. Yet, they were lucky enough to crash on a small pastoral world, a lush green planet whose surface comprised a stunning array of different landscapes. Rolling plains, grassy, undulating hills, and swamplands under the cloud cover of the perpetually gray twilight side of the planet.

Naboo.

The human colonists from Grizmallt ventured to Naboo's Gallo Mountains, and settled a farming community at the Dee'ja Peak.

Colonizing other planets in the Star Wars galaxy seems to have been a piece of (moist) Gungan cake. No demand for bio-domes. No call for acclimatization. And no need for breathing devices. You simply crash landed and, bingo, before you know it you're up and running a farmstead on a perfect-looking planet. You can even eat the local delicacies.

Is life in the Star Wars galaxy really that simple and straightforward? How come so many planets have such clean air, and breathable atmospheres? What's going on?

How Breathable Planets Are Made

The breathable atmosphere of the Earth today is influenced by a number of factors.

Gravity, sunlight, seas, topography, and terrestrial biota (Earth's animal and plant life). Some of these factors are local, others more global. With so many intricate influences, it's easy to see why the weather is hard to forecast in some places. It's also easy to see why an atmosphere's composition—the "air" of other worlds if you like—could be so very different to our own human-breathable air.

The story of air starts way back, both for our own planet, and the exotic worlds in the Star Wars galaxy.

The original Earth of 4.6 billion years ago would have been mostly particulate, slowly forming out of the gaseous nebula swarming about the primordial Sun. In time, the gases condensed into liquid and solid states. Some cooled to start the seas, some became the more compacted continents. But the Earth's core continued to rage with a ferocious heat that makes our planet alive.

Above all this, like a soccus around the surface of this abundant living sphere, sits the air.

Scientists believe the original atmosphere of the primal Earth escaped into space. Compared to today's breathable air, this primary atmosphere was toxic—rich in ammonia, neon, water vapor, and methane. The air was without breathable oxygen. Nonetheless, unicellular organisms arose and then breathed out oxygen into the air, which brought about a revolutionary change in the Earth's atmosphere. Over many millennia, this atmosphere then evolved into the air that we breathe today.

Wherever planet production occurs, the same process is played out. Air evolves.

Knowing that, we might expect the air of a pastoral planet like Naboo to be at least a little different to the atmosphere of a desert world like Jakku, and in contrast to an icy planet like Hoth—different topography, different biota, and possibly different gravity. Crucially, there'll also be a varied level of sunlight on each world, depending on the relative distance from the parent star. Oceans will play a part, too. Only 8% of Endor's surface was covered with water, compared with 71% of Earth's, and Kamino—the Star Wars world where the clone army for the Galactic Republic was created—was an entirely aquatic planet.

Planetary Stats

Yet, Star Wars doesn't imply that *all* planets in its galaxy have human-breathable atmospheres.

We simply don't see the huge number (in fact the overwhelming majority) of planets that have poisonous air. Humans like Han and Leia don't visit these wayward worlds, as human civilization would not have developed there in the first place, unless the planet had been terraformed. If Naboo hadn't been breathable, its human colonists would have had no tale to tell. They would simply have perished in the toxic air, and would have left the planet to the Gungans. However, we do also catch the odd glimpse of un-breathable worlds, such as the gas giants Endor and Yavin, which we spy from space.

But how many planets are there in the Star Wars galaxy to examine?

We can again use our own Milky Way Galaxy as a guide. Our Galaxy contains up to four hundred billion stars, and experts believe that in orbit about these stars sit at least one hundred billion planets, many of which are Earth-like. There are galaxies more gargantuan than the Milky Way. Our near neighbor, the Andromeda Galaxy, holds as many as one trillion stars, and a proportionate number of planets. Canon quotes the dimensions of the Star Wars galaxy merely as being "over 100,000 light-years" in diameter. The Milky Way could be anything up to 180,000 light-years across, and Andromeda is approximately one-fifth larger than that. It's likely, though, that the Star Wars galaxy is purposely on a par with our own, in terms of stars and planets.

So, in a galaxy perhaps numbering over one hundred billion planets, a few dozen named and breathable worlds doesn't seem to be such an exaggeration.

Galactic Habitable Zones

The Star Wars canon suggests that its human population began with the Coruscanti.

Their home world sat near the center of the galaxy, a perfect place from which to explore that part of the galaxy's stellar population that is most densely packed. In our Milky Way, the Galactic core is five hundred times more densely packed with stars than the neighborhood of our Sun, which is about 26,000 light-years from the center of our Galaxy. Whether the Coruscanti would have a free choice of planets to pick from, within the parameters of their galaxy, is a matter of some debate.

Some scientists have suggested the idea of a galactic habitable zone (GHZ).

This habitable zone is the region of a galaxy in which life is most likely to develop. In particular, the notion of a GHZ includes factors such as the rate of potentially life-threatening major cosmic events, like supernovae, and an idea known as metallicity—the fraction of chemical elements that is not hydrogen and helium. These factors help in the experts' calculation of which regions of the Galaxy are best suited to forming Earth-like planets, or to evolve suitable environments for life to advance.

And yet, more recently, there are scientists who believe that the GHZ may extend to the entire Galaxy, rather than habitability being restricted to a specific region in space and time.

The Colonizing Coruscanti

Out of their homeworld the Coruscanti came, in spaceships sleek and striving.

As Coruscant was a Core World, these early humans were fortunate to find their part of the Star Wars galaxy replete with other stars. And those stars orbited many other worlds. Of those planets, what number would have been habitable?

On November 4, 2013, based on *Kepler* space mission data, scientists boasted there could be as many as forty billion Earth-sized planets orbiting in habitable zones in the Milky Way Galaxy. Some estimates suggest there are around five hundred Sun-like stars within one hundred light-years of our solar system, and a typical area of a galactic core is five hundred times more densely packed than that. The Core region of the Star Wars galaxy would be crammed with potential candidate planets for colonization.

Over millennia, if we assume the Coruscanti had sophisticated spacecraft, they could sample any candidate planets from afar, using any on-board remote sensing kit at their disposal. Many worlds, of course, would already have been inhabited by a cornucopia of other creatures. Yet more worlds—those found to be toxic to humans or that had strategic potential within a developing galactic network of worlds—could be terraformed.

Terraforming literally means "Earth-shaping," or in the case of Star Wars, we might call it "Coruscant-converting." In either case, it's the process by which a moon, planet, or other body is calculatingly controlled in terms of its temperature, atmosphere, topography, and ecology to evolve into an analogue of the Earth's environment—or Coruscant—to make it habitable by humans or Earth-like life.

That Mos Eisley Cantina

Star Wars suggests it may be possible to terraform worlds without the loss of indigenous life.

With the clear and apparent level of cultured technology in the Star Wars galaxy, the implication is that terraforming a planet for human habitation may be possible without disrupting species native to the world being terraformed. Terrestrial scientists doubt this, naturally. Sure, they agree that terraforming may be possible, transforming hostile planets into habitable ones by somewhat less-than-sophisticated means. But Earth scientists draw the line in believing it may be possible to terraform Tatooine without strangling the Sand People.

This seems like a good point to invoke the "Three Laws of Arthur C. Clarke." The British science fiction writer once suggested these three adages when pondering predictions:

Clarke's First Law:

"When a distinguished but elderly scientist states that something is possible, he is almost certainly right. When he states that something is impossible, he is very probably wrong."

Clarke's Second Law:

"The only way of discovering the limits of the possible is to venture a little way past them into the impossible."

Clarke's Third Law:

"Any sufficiently advanced technology is indistinguishable from magic."

So with that Mos Eisley cantina in mind, and a mischievous glint in the eye, we may conclude that there is some superior terraforming technology being used in the Star Wars galaxy, which is a venture into the impossible, indistinguishable from magic.

WOULD DNA BE THE REPLICATOR IN THE STAR WARS GALAXY?

What was the genetic makeup of a Wookiee? What kind of blood coursed through the veins of Yoda? What was the evolutionary history of the Hutts?

The laws of physics are true for all corners of the cosmos. Planets, stars, and galaxies come and go in the same way they would in our solar neighborhood. But what principles of biology on Earth would also have stood in Star Wars? What of biology is particular to Earth, and what applies elsewhere in the universe?

When looking back at the alien life-forms in the Star Wars galaxy, we find customary creatures, many humanoid in form. We also see much that is almost too unearthly to imagine. But is there anything about life in the universe that is true for all life? Even if life elsewhere is based on silicon, rather than carbon, and uses a solvent such as ammonia rather than water, is there a common commodity to life, which is far more fundamental?

Evolution

Evolution itself is lifeblood.

It's the perfect theory to understand the history of Star Wars species, as the rise of evolution injected a sense of history into our science. "He who . . . does not admit how vast have been the past periods of time may at once close this volume," Darwin wrote in the *Origin*. For species to have evolved on Earth, or in Star Wars, the genuine extent of history had to be much longer than the roughly six thousand years suggested by some in Darwin's day. While Victorian biology and geology implied the Earth was ancient, they did not prove it.

But Darwinian evolution is the historical cipher to understanding the physics and cosmology of Star Wars, too.

Evolution in biology begat other evolutions. The spirit of Darwinism was gifted to physicists. They, too, approached the question of the age of our planet, and the age of the Sun. First, they used thermodynamics, the branch of physics concerned with the dynamics of heat energy. Then, late in the nineteenth century, the nuclear age dawned. Radiometric dating—the technique for dating materials using naturally occurring radioactive isotopes—provided age-dating to fields as diverse as geology, astrophysics, and cosmology. We know now that the planetary systems of the Star Wars galaxy would have gone through similar evolutions.

And so, the philosophy of evolutionary change is key to understanding the nature of the cosmos.

The history of the Star Wars galaxy must have been a long one, as our history has been revealed by evolution to be not only complex, but *big*. Once we realized the Earth was old, and that all had been fashioned by lengthy processes of change, we began to see that the story of our planet was part of an even older tale. And the materialist bedrock of evolution has influenced our perception of all things: culture, language, the society in which we live, and every single discipline of science itself. So it would have been with Star Wars.

Everywhere we looked in Star Wars, we would have seen systems of matter in motion, evolutionary change. The swirling disc of gas and dust in the early stellar nebulae gave birth to the galaxy's stellar systems—the planets growing out of the nebulae, evolving as they circled the young suns, such as Jakku, Ileenium, and the Endors. Electron clouds constantly swarmed the heart of the galaxy's hydrogen and helium atoms, which brewed within stellar interiors into heavier elements, gifting to the galaxy an evolving chemistry.

On Earth, evolution caught up with cosmology, too.

When Einstein had composed his general theory of relativity, he'd assumed that the universe was static. Before long, telescopic redshift evidence from the likes of Edwin Hubble told the telltale signs of an expanding and evolving space-time. The small, static, and Earth-centered cosmos demolished by Galileo has given way to the notion of an evolving universe so large that light from its outer reaches takes longer than twice the age of the Earth to reach our telescopes, and galaxies abound.

Replication

So, our evolving conception of the cosmos is changing as rapidly as the universe itself.

Have we yet learnt enough to say what aspect of life elsewhere would be common with life on Earth? In considering this question, scientists have decided that common factor would most likely be replication. On Earth, all life evolves by the differential survival of duplicating entities, which we may call replicators. On our home planet, the replicator that does the job is the DNA molecule.

DNA has been called the most extraordinary molecule on Earth.

DNA, or "deoxyribonucleic acid," exists simply to make more DNA. There's nearly two meters of DNA squeezed into almost every human cell. Each length of DNA has about 3.2 billion letters of coding. That's enough to enable $10^{3,480,000,000}$ potential combinations. That's a huge number of possibilities. If something like DNA existed in the Star Wars galaxy—and we will soon argue that it would have—the possibilities in a replicator like DNA means the potential people who could have been in the place of Luke Skywalker, but who never saw the light of Tatoo, outnumber the sand grains of Jakku. Such unborn ghosts include greater philosophers than Yoda, pilots greater than Han Solo. That's because the parade of possible people allowed by replicators like DNA so hugely outnumbers the parade of actual people.

Imagine Princess Leia, standing in front of a mirror. If her biology works anything like ours, she could reflect on the fact that she is gazing at ten thousand trillion cells. Almost every one of them contains two yards of jam-packed DNA, or some other replicator. If it were all spun into a single thread, it would make a solitary strand long enough to reach the Moon and back, many times over. Humans have as much as twenty million kilometers of DNA, scrunched up inside.

Humans are vehicles for DNA.

We are hosts for DNA parasites, which are our genes. In many ways, the actual function of life, that which is being played to the max in habitats on our home planet, is DNA survival. However, DNA is not floating free; it is locked up in living bodies. And though we could not live without it, DNA itself is not alive. In fact, it's among the most chemically inert molecules in the natural world. That's why it survives in the murder room of the forensic scientist, and can be teased out of the prehistoric bones of a Celtic warrior. Curious that something so lifeless should be at the very core of life on Earth.

The Cipher of Life

The river of DNA flows through time.

It is a river of coded information, rather than blood and bones. Much of the DNA is redundant, about 97% of it, in fact. But the genes, the business end of DNA, are the short sections that control and orchestrate vital functions. Genes have been compared to the keys of a piano, each playing only a single note. And the combination of genes, like the combination of piano keys, creates chords and a vast variety of tunes. Coalescing, the genes' contributions create the orchestration that is the human genome.

The human genome is considered to be a kind of instruction manual for the body. Genes are simply instructions to make proteins. The letters in which the instructions are written are known as bases, and they are the key to the genetic alphabet. The bases consist of four nucleotides: adenine, thymine, guanine, and cytosine.

What would the cipher have been in the Star Wars galaxy? What ingredients would do the job of replicator?

Lab results on Earth are promising for alien ciphers. Experiments show that it's possible to tweak practically any feature of our DNA system. In other words, even though the roles that the molecules play may be universal, the actual molecules used on Earth are very provincial.

Not only do alternatives to DNA exist, but scientists also believe it's possible to extend the range of bases. It's achievable to add new letters to the genetic code. In fact, these new bases can be inserted into DNA, increasing its alphabet from four letters to twenty-four! The resulting replicator, what we might call "Star Wars DNA," could create novel and quite different proteins.

We have our answer. The specific components of our pervasive DNA/protein chemistry are quite parochial. The chemistry can be changed at most levels of the process, without messing with the basic activity of information-carrying and replication of life. There's nothing sacred about DNA and the genetic code.

Star Wars Chemistry

Star Wars biochemistry would have been different in most details from today's earthly counterpart.

However, even in the unlikely prospect that the two biochemistries were precisely the same, the cornucopia of creatures on Earth would not have their duplicates on Dantooine, Kamino, Sullust, or anywhere else in the Star Wars galaxy. This would be true even on another Earth-like planet, for reasons of evolution and life's diverging pathway on each. In evolution, context is everything.

That's why there would be no dinosaurs on D'Qar, and no Neanderthals on Naboo.

Not unless they were transported from Earth, but that sounds like a plot from *Episode XXXVII.* The beginning of life on Star Wars planets, however, may have a lot in common with Earth, especially if those planets are also aqueous: an initial bombardment by comets and asteroids, and broiling oceans like a turbulent biochemical soup, would make them all very alike.

DNA clearly worked on Earth. So anything sufficiently similar would have worked in the Star Wars galaxy or any other. Given what we now know about alternatives to replication on Earth, it would be folly indeed to think that DNA was the only way to replicate.

DARWINIAN DYNAMICS: HOW DO THEORIES OF EVOLUTION APPLY IN THE STAR WARS UNIVERSE?

If there's one image from *The Force Awakens* that conjures up thoughts of Darwinian adaptation, it's Rey's struggle to survive in the harsh environment of Jakku, with no family to support her.

Yet, since the days of Darwin, scientific research in biology has shown that evolution is more complex than mere adaptation. It occurs both above and below this conventional Darwinian level of the adaptive struggle of species for survival. It's not all just Hutts with blasters.

Below the conventional level is the random mutation of genetic material, leading to unsolicited and serendipitous change. Above the conventional level is the prospect of impact events, the kind of cosmic catastrophes that have punctuated life's pathway on Earth, and are used to mark geological time itself. We humans, on our home planet, are prey to all three levels of evolutionary change. Biology is admittedly parochial, as there's only one Earth to study, but if we assume evolution is universal, there is little reason to doubt that the Star Wars galaxy would also be subject to these levels of evolutionary change.

Natural Selection 101

Darwin himself first set out the case for evolution in his book *The Origin of Species*.

The theory comprised three component concepts. The first notion was "variation." Variation is based on the observation that each and every individual of any particular species is different. Wookiees may all *look* the same, but each has a unique set of quirks and characteristics.

The second notion is "multiplicity." Living creatures, including alien species, tend to make more offspring and have bigger broods than the environment can necessarily maintain. "Nature," according to Alfred, Lord Tennyson, "[is] red in tooth and claw"—both in our Galaxy, and, we may assume, in galaxies far, far away. One might live in a world in which only a small fraction of the Ewoks and Exogorths that are brought into existence actually survive, or manage to evade predators long enough to mate.

The third notion was natural selection. The individual differences between members of a species, coupled with the environmental forces, shape the likelihood that a particular individual will last long enough to pass its characteristics on to posterity. By the survival of the "fittest," Darwin didn't mean some kind of innate superiority to others. Rather, it is an unsolicited advantage, the result of the pure serendipity of freak change; they just happen to better "fit" their environment. Should the scene change in time, those previously well-adapted may swiftly discover they no longer fit. Soon, other freaks may inherit the galaxy, as it were.

Cedar Forest from the Trees?

Welcome to the world of constant change. Whether it's Earth or Endor, natural selection is the engine that creates new species. Nature favors variety, and prefers geographical spread. The further afield a species is scattered, the less tied their prospects are to a single setting. So what about those Ewoks stuck in their famous forests? They clearly evolved from a line of successful ancestors, as they have survived to tell the tale. But how might environmental forces have shaped their development? How well do they fit their imagined habitat?

Let's consider the Ewok in more depth. This diminutive species of furry upright bipeds stood at an average height of about one meter tall. The presence of an opposable thumb enabled them to use tools as well as weapons such as spears and slings. Although Ewoks were quick learners, there appears to have been some element of arrested development. When discovered by the Empire, they had yet to progress past Stone Age-level technology. Nonetheless, they were skilled in forest survival and the construction of primitive technology, even using hang gliders as vehicles.

Ewoks had evolved to a hunter-gatherer stage of evolution. They had pottery, hunting, and primitive flight, along with the procurement of fire. They spent most of their time in the treetops, in villages built between the closely spaced trees, or venturing out onto the forest floor to forage and hunt.

Could environmental forces explain Ewok arrested development?

A key factor in human progress on Earth was the mastery of metal. The invention of metal smelting, the alloying of copper with tin to usher in the Bronze Age civilizations, and the eventual transformation into the Iron Age, were critical turning points in human evolution. The use of charcoal was a huge breakthrough. Charcoal burns at 1,100°C, and that was enough to melt metal out of rock.

Stone Age Sanctuary Moon

The age of metals does not seem to have started on Endor. Most metals, apart from gold and tiny amounts of copper, are not found in their raw state. That's why ancient humans first used miniscule amounts of gold and copper for ornaments and jewelry only. There's no sign of such adornment on the Ewoks. It's all twine and leather. The state of their pristine forests suggests they've not smelted metals using charcoal either. In prehistoric times, Britain had been almost completely covered in woods. Yet, by the end of the 1500s, 90% of the ancient forests had gone, sacrificed for the sake of metallurgy. Ewok trees still stand.

Maybe there are no metals on the Sanctuary Moon.

However, if the Endor system is anything like the solar system, metals are more than likely. When stars like the Sun and Endor I and II were made, the high temperatures of the inner protoplanets were too hot to hold the volatile gases that dominated the stellar nebula. Only high-melting-point materials, like iron and rocky silicates, were stable. Thus, rocky planets are made primarily of metallic cores and silicate mantles with thin—or absent—atmospheres. Although the gas giant Endor is in the outer part of the Endor system, its satellites—such as the Sanctuary Moon—are also likely to be differentiated to iron and rocky silicates.

Take our own planet/satellite system. The percentage of concentration of metallic elements on our Moon compares very favorably with the Earth. Indeed, the lunar highlands and lowlands have more iron than Earth, and the Moon is similar in composition to the Earth's mantle. But have there been events on Endor that would have accreted metals into mineable concentrations? Ores can become concentrated by hydrothermal processes, magmatic processes, or big-impact events. Of these three, the latter two are the more likely on Endor. It could be that the first Imperial outpost on Endor discovered some precious metals, but in too low of concentrations to be of any use.

Cosmic Catastrophe on Coruscant?

All things considered, it seems Stone Age Ewoks makes some sense.

What of impact events in the Star Wars galaxy? What news of the kind of cosmic catastrophes that have been ubiquitous on Earth, and elsewhere in the solar system? The Earth is about 4.5 billion years old. The earliest undisputed life on Earth dates from around 3.5 billion years ago. Yet, there's physical evidence of life in biogenic graphite, which is 3.7 billion years old, as well as the remains of biotic life found in rocks 4.1 billion years old. Life on Earth clearly didn't really get started for around half a billion years, as a result of bombardment of large amounts of cosmic debris during the coalescence of the solar system.

Even after life had found a foothold on Earth, cosmic catastrophes from comets and asteroids played a major role in our history. Impacts may have been an important factor in a dozen or so mass extinction events recorded in the last half-billion years alone. Even looking to the future, it is clear that impact events on Earth will continue to occur.

Comets exist outside the solar system, too. Such exocomets orbit stars other than the Sun. The first exocomets were detected in 1987 around Beta Pictoris, a very young star. To date, in our Galaxy eleven stars have been confirmed as having observed or suspected exocomets in orbit around them.

But were there comparable comets and catastrophes on Coruscant, for example?

In all likelihood, yes. During the Clone Wars, a Republican detail accompanied by a D-Squad of astromech droids encountered a comet storm near the planet Abafar, a remote desert planet located in the Outer Rim Territories. More tellingly, aeons before the Battle of Endor, a comet named Kinro was predicted to destroy several Core planets. The comet was destroyed before it reached the Mid Rim, and its destruction was attributed to the Jedi Order, who banded together and willed the comet to break apart. Several Jedi lost their lives or minds as the dramatic events unfolded.

Yet, there's something amiss with this account. The Core Worlds (also known as the Coruscant Core) include the likes of Alderaan (whose fate was sealed anyhow), Corellia, Hosnian Prime, and Coruscant itself. But the Mid Rim is a region of the Star Wars galaxy located between the Expansion Region and the Outer Rim Territories. In other words, the Mid Rim is light-years—to say the least—away from the Core Worlds. Is it possible to have a comet that moves between stars, as suggested by the Star Wars canon?

It seems so. Although no interstellar comet not gravitationally bound to a star has been conclusively classified, such comets are thought to exist. They're ejected from their home systems, potentially by gravitational scattering from nearby planets, or perhaps by passing stars. Kinro would have to have experienced some serious interstellar forces to make its way from beyond the Mid Rim to where it was destroyed, so let's assume dark forces were involved.

The Star Wars galaxy clearly has been subject to different levels of evolutionary change. If we open our minds to take a deeper look, we can make some scientific sense of what we see. As for random mutation of genetic material in Star Wars, and the prospect of midi-chlorians, these are topics for entries elsewhere in this book.

HOW ARE THERE HUMANS IN THE STAR WARS GALAXY?

Humans. There's no escaping them in the Star Wars galaxy. They're everywhere: Naboo, Alderaan, Tatooine. But the question arises: How did they get there?

Canon has it that the human species are native to the planet Coruscant. This presents a few problems for our own theories of human evolution, considering that Star Wars took place a long time ago, in a galaxy far, far away.

For a start, we can identify a steady process of change and adaptation that has led to the many species currently present on Earth. We have a fossil record of extinct organisms that spans almost four billion years of geological time. Then there's the genetic evidence held within the DNA of extant as well as ancient species.

So what's the deal? Could we be somehow related to humans in a distant galaxy, or could they represent something even more absurd, an independently evolved species identical to humans?

Straight Out of Coruscant

The Star Wars galaxy is well-populated with humans—the natives of the city planet Coruscant. From this home near the core of the Star Wars galaxy, they migrated out to colonize the many star systems that they now have a presence on. A similar thing is thought to have happened here on Earth, only involving continents rather than star systems.

The mainstream view is that modern humans evolved in Africa then spread to colonize the other continents. These modern humans replaced the earlier hominins such as *Homo erectus* and Neanderthals that had already started to populate parts of the world. It's called the Out of Africa Hypothesis.

Another theory of modern human origins is the multiregional hypothesis proposed by Milford Wolpoff, Alan Thorne, and Xinzhi Wu. In this model, modern humans evolved locally from the earlier hominins that were already occupying different parts of the world, having left Africa many years before. There is not as much support for the multiregional idea, though.

However, if a multiregional evolution of humans had occurred in the Star Wars galaxy, it would mean that an early migration of humans propagated to occupy various star systems, then evolved to become the modern humans that we see living on the different planets. We'd expect to see more variation in the humans than we do, though. As such, the humans of Star Wars probably populated the galaxy following an "Out of Coruscant" model.

But what about us earthbound humans? Where do we fit in? Is it possible that we, or possibly life on Earth, could have originated from the Star Wars galaxy?

Evolving Humans

The idea that life on Earth could have been seeded from somewhere off the planet is called the panspermia hypothesis. The idea has arisen a few times since the fifth century BC. The most recent advocates of the theory were Sir Fred Hoyle and Professor Chandra Wickramasinghe.

According to Wickramasinghe, panspermia is the hypothesis that life exists throughout the universe, distributed by meteoroids, asteroids, comets, and planetoids.

The main thrust of this theory isn't concerned with advanced life-forms, though. It mainly deals with microscopic organisms such as viruses, bacteria, and Achaea as well as molecules that can act as the building blocks of life. Additionally, Wickramasinghe says that it "is not meant to address how life began, just the propagation method that may cause its sustenance."

Since its initial appearance, life has gradually diverged to become the many different species that have occupied Earth. Even before we knew what genes were, we could see common features and traits between past and present life-forms on Earth. The degree of relatedness between these life-forms is often represented as a tree of life, the branches of which are formed by the process of evolution.

On our earthborn tree of life, we have various degrees of relation such as primate, mammal, and vertebrate. However, in Star Wars there don't appear to be many animals related to a human evolutionary past. In fact, Coruscant is a huge, planet-spanning city, or *ecumenopolis*, seemingly devoid of any natural formations. On Coruscant it would seem that nature has long left the building, with the human population regarded as the only true natives.

So how could humans evolve on Coruscant and yet be the only natives on the planet, considering we have so many species native to Earth that are still in abundance on Earth?

It's possible that the humans on Coruscant evolved to out-compete the other native life-forms, eventually overrunning and urbanizing the whole planet. However, the likelihood of them being genetically similar to humans is impossible, given the huge number of random events that took place on Earth to allow humans to exist in our current form. For Coruscanti to be the same species as us, we would have to occupy the same branch on the tree of life.

Our particular human branch, *Homo sapiens*, diverged from other hominins about two hundred thousand years ago. Assuming this to be a reliable explanation for the origin of humans on Earth, sometime in the last two hundred thousand years the human species managed to end up in two different galaxies. Now, as there is more evidence for an earthbound evolution of humans, the question is: How has the human genome managed to propagate to the Star Wars galaxy from Earth?

So, let's recap: It's extremely unlikely that two independent evolutionary paths would lead to the same identical species, so the humans on Earth and Coruscant would have to be related. There's a lot of evidence for human evolution on Earth, but there's little evidence of humans evolving on Coruscant. It seems more likely that humans evolved and originated from Earth.

How Could Humans Also Be in the Star Wars Galaxy?

Regarding the aforementioned panspermia, it mainly considers microscopic seeds of life propagated by space rocks. However, in 1972, Francis Crick and Leslie Orgel considered directed panspermia. This theory considers the possibility that organisms were deliberately transmitted to Earth by intelligent beings from another planet.

To extend that idea, could similar intelligent beings have taken an already evolved human species from Earth to Coruscant?

When we first saw Star Wars, the film took place "a long time ago in a galaxy far, far away." Exactly when is never identified but a long time ago could be as little as a thousand years to as large as many billions of years ago. Considering that modern humans only appeared about two hundred thousand years ago, that would be the earliest point that the human species could have left Earth. Therefore, "a long time ago" was anytime after that time.

The documented history of the Star Wars galaxy goes back thousands of years. Millennia before the battle of Yavin (the battle that blew up the first Death Star), an ancestral species unlocked the secrets of hyperspace. They were apparently inspired by observing massive whale-like creatures called Purrgil that have the ability to travel through hyperspace.

Hyperspace travel is a given in the Star Wars galaxy. Journeys are regularly made almost halfway across the galaxy in a matter of hours, days, or, at the very most, weeks. Putting notions of a hyperspace disturbance aside, what's stopping an ancient intelligent civilization from deciding to leave the galaxy altogether?

Intergalactic Seeding

Let's imagine a conservative journey time of four weeks to travel one hundred thousand light-years (the distance across the Star Wars galaxy). In one year that would mean a journey of 1.3 million light-years could theoretically be made; given that fuel was not an obstacle to these beings.

Then in one hundred years, they could potentially reach a galaxy that's up to 130 million light-years away, i.e., our Galaxy.

For reference, there are about 2,500 large galaxies and fifty thousand dwarf galaxies within one hundred million light-years of the Milky Way.

Now, depending on whether these intelligent aliens have a hands-on or hands-off approach, there are two options for the humans they kidnap from Earth. They could use carbonite freezer technology to keep the humans in stasis for the long journey, or they could let the humans live and breed on a generation starship.

The hands-off approach would require fewer resources; however, the humans would arrive on Coruscant with a primitive knowledge of how to manipulate their surroundings. They could then be left to migrate across the planet with the occasional technological assist from the intelligent alien propagator.

A more hands-on approach could see the humans taught how to use and create advanced technology. Generations would grow on the ship, learning in a similar way to how we do on Earth. They would start off knowing nothing, then be educated before joining the work force and subsequently inventing their own technologies. All humans start as a blank state before becoming capable actors in the world.

When these humans arrive at Coruscant it could still be almost two hundred thousand years before our present day. With fifty thousand years of technology and breeding they could easily achieve the 680 billion humans that populate Coruscant, and that's even with loss from outward migration to other planets.

COULD WE EVER BECOME JEDI?

According to canon, before the dark times and the empire, the Jedi Knights were the guardians of peace and justice in the old republic. They acted sort of like an interstellar United Nations. At their peak, about ten thousand Jedi kept watch over the whole galaxy. They did so for over a thousand generations.

Jedi can channel the Force to read or manipulate people's thoughts, move objects at a distance, and even sense the future. Their "sorcerers' ways" also mark them out as extraordinary amongst other members of their species, while their Order was often seen as a form of religion.

Not anybody can become a Jedi. It isn't something you're born with either, although having a strong lineage can help. To become a Jedi, dedicated training and discipline are vital.

Although the Jedi are a fictional creation, there's been a surge in the number of people entering Jedi Knights as their religion on the national census of their country. Many may have done it for fun, but some people do actually try to live by the Jedi Code. Could we ever have real-life Jedi, though?

The Jedi Traits

What does it take to be a Jedi?

The original trilogy showed us that Jedi could be humans as well as Yoda's unknown species. Since then we've been able to see just how inclusive the Jedi order is, with representative species from all over the Star Wars galaxy.

Species range from Anxes to Zabraks. There's Admiral Ackbar's amphibious species, the Mon Calamari, and the serpentine Thisspiasians. Some species, like the Quermians, even had two brains while others like the Besalisks had four arms.

Despite their different body plans, what they all have in common is the ability to recognize and make use of the Force. Beings with this ability were called "Force-sensitive." This skill was often present from birth. Force babies were of special interest to the Jedi, who kept a database on their locations across the galaxy.

Potential Jedi could further be identified by the presence of particular skills and abilities, described as Jedi traits. This could be above average intelligence or quick reflexes, both of which the young Anakin Skywalker displayed. These traits aren't limited to the Star Wars galaxy, though; Earth is full of such marvels.

The Force Is Strong

With a human population on Earth of close to seven billion, and 350,000 babies born every day, some people are bound to develop outstanding talents. Obvious examples are individuals with particular aptitudes for arithmetic, art, or athletics. Many of these people exhibit a talent at an early age, excelling in particular disciplines at school.

Talented youth are often head-hunted for sports academies, while organizations like Mensa have established gifted and talented support programs for particularly clever youth. Mensa's core aim is to identify and foster human intelligence. There are more than 57,000 Mensans across the US, ranging from age 2 to 102.

When Anakin was discovered, he was already building robots and podracers. As seen in *The Phantom Menace*, young Skywalker was eager to operate his racer in competition. Imagine a nine-year-old kid building and racing motorbikes on the weekends. His skills were recognized by Qui-Gon Jinn, who remarked, "The Force is unusually strong with him." Through the Force, Anakin could see things before they happened, which is why he had such quick reflexes.

Quick reflexes and reaction times rely on a person responding to stimuli. This depends on their cognitive processing speed. There are many humans who rely on quick reflexes, particularly in sport. Some people, such as Japanese swordsman Isao Machii, exhibit such quick responses that it has led psychology professors to describe his ability as processing on an entirely different sensory level.

So superhumans do exist. However, in Star Wars, it is sensitivity to the Force that sets them apart from the average person. By channeling the Force, they can become aware of stimuli that normal humans are not. In *The Phantom Menace*, for better or for worse, we found out that the will of the Force is communicated through what are called "midi-chlorians."

Midi-Chlorians

According to Qui-Gon Jinn, "Midi-chlorians are a microscopic life-form that reside within all living cells and communicate with the Force. . . . We are symbionts with them. . . . Without the midi-chlorians, life could not exist, and we would have no knowledge of the Force."

Although these midi-chlorians are present within all life-forms according to Qui-Gon Jinn, they sadly aren't present in us. However, when George Lucas devised midi-chlorians, he was inspired by tiny organelles that really do exist in nearly all the cells of our bodies. They're called mitochondria.

Mitochondria act as the power plants for the cells in our bodies. The more mitochondria a person has, the better their endurance performance will be. They are the basis for oxygen-breathing life-forms, generating energy from food. Without them, the majority of life as we know it would not exist.

We inherit our mitochondria from our mothers. If midi-chlorians are inherited like mitochondria, then Anakin would have inherited his midi-chlorians from his mother. However, she never exhibited signs of Force sensitivity.

The converse argument—that midi-chlorians are passed down by the fathers—falls apart, too, when we consider Kylo Ren's abilities. It's possible that midi-chlorians represent a dominant genetic trait, and whoever is stronger with the Force passes on midi-chlorians to their offspring.

In Anakin's case, rumor has it that he was conceived from the Force itself. This supports the realization that Anakin's midi-chlorian count had a reading that was off the charts. However, there might be another explanation.

Mitochondria multiply independently of the cell cycle, in response to energy needs. So the more exercise that a person does, the more mitochondria are produced in their cells. They essentially respond to demand. If midi-chlorians act in a similar way, then their amount could also be stimulated by the demand placed on them, i.e., they could increase with exposure to the Force.

Jedi Training

To become a Jedi Knight takes ability and training. There are parallels between their journey and that of the real knights who roamed medieval Europe.

Potential Jedi are found and trained from an early age based on their Force sensitivity, whereas young hopefuls in real life had to be of noble descent. At age seven, they could become a page and were looked after by the lords of the castle. This is equivalent to being taken to the Jedi Temple on Coruscant to become a Jedi youngling.

Jedi younglings would go through "The Gathering" on the arctic planet of Ilum, where they would seek out the kyber crystals that would become the heart of their first lightsaber. After completing their Initiate Trials, they could then become a Padawan to a Jedi Knight.

In medieval times, pages would accompany knights into battle, learning from them and becoming a squire at age fifteen. This was also when they could get their own armor.

At age twenty-one, a squire could be knighted in an accolade ceremony, whereas a Padawan would become a Jedi Knight after successfully completing their Jedi trials. Jedi Knights followed the Jedi Code, while real knights had a similar chivalric code.

Knights were often revered, and to this day people are still knighted in the UK, although with a different emphasis than with medieval knights, who were trained soldiers.

There was even a religious order of knights called the Knights Templar who were tasked with the protection of Christian pilgrims. The head honcho was called the Grand Master, a term also applied to the top Jedi, Grand Master Yoda.

It's no surprise that the Jedi ranks have parallels to real-life practices. Fact often informs fiction. Maybe a more interesting scenario is how a fictional following can become the fact.

May the Force Be With You

The Jedi are an exclusive organization of Force-sensitive individuals, but when we eliminate the element of the Force, we see how similar they are to individuals in reality.

They represent a hope of achieving something beyond ourselves, whether it's superhuman powers, or a connection with some entity that's bigger than us yet within us all.

George Lucas said, "I put the Force into the movies in order to try to awaken a certain kind of spirituality in young people. More a belief in God than a belief in any particular, you know, religious system."

In 2001, more than 390,000 people signed their religion as Jedi in the UK national census. As of the 2011 census, England had the biggest number of Jedi in the world, numbering 176,632.

In response to this, an Anglican bishop in the UK said that a movement would have to be functioning for a good amount of decades before it could be considered as a serious religion.

We may not be able to be Jedi by nature, but it's possible that one day we may see official Jedi by name. There's still hope.

WHY ARE WOOKIEES SO MUCH HAIRIER THAN HUMANS?

It's the huge, hairy, huggable buddy of Han Solo, with the face of a hound and vocal dexterity of a black bear. He has Han Solo's back in any venture or scrape, but he's much more than just Han's best friend. He's Chewbacca, the two-hundred-year old Wookiee.

Based on George Lucas's pet dog Indiana, this walking, "talking," brawling teddy bear entered our imaginations, right alongside vague notions of Sasquatch and Bigfoot. However, he looks more like the Star Wars prediction of what would have happened if an early ancestor of dogs had evolved to walk on two legs, rather than our primate predecessors.

Considering that humans evolved from a hairy ancestor and gradually lost our hair after becoming bipedal walkers, why wouldn't the same thing happen to Wookiees?

Covered in Hair

Most mammals have a visible layer of thick body hair, which they use as insulation to keep in their body heat. However, there are a few that don't, including elephants, pigs, and naked mole-rats.

Aquatic mammals like whales evolved thick layers of fat to help regulate their body temperature, while bulkier mammals like elephants take longer to heat up and cool down, making them more resilient to environmental temperature changes. However, in colder climates, thick hair was still of value, as seen in woolly mammoths.

The adult human body has about five million hair follicles, distributed quite evenly across most of the body. It's a similar number and density to our nearest primate, the chimpanzee. The major difference is in the thickness, length, and darkness of the hairs.

Most of our hair follicles produce small, unpigmented vellus hairs rather than thicker, pigmented terminal hairs such as the hair on our scalp. This makes us appear to be quite naked despite being covered in hair.

During human adolescence, the vellus hairs typically become terminal hairs in the armpit and pubic regions. However, with a rare few, this change can occur to the vellus hairs on other parts of the body. It's known as hypertrichosis, sometimes referred to as werewolf syndrome.

So it's possible that Wookiees are a species afflicted with werewolf syndrome, unless they're the norm and it's just us who are the oddballs?

Hair and Survival

It's believed that bipedal walking was a necessary precursor to our transition from a dense to sparse hair covering. This transition happened more than one million years ago on the African savanna, with our *Homo erectus* ancestors.

Scientists Graeme Ruxton and David Wilkinson highlight that for humans developing in the hot, open African savanna, physical activity was less heat-exhausting during the cooler parts of the day. As hair loss increased and sweating improved, humans were able to extend the time they were active during the day. This gave them an advantage for obtaining food, water, and anything else they needed to survive.

Kashyyyk is the temperate home planet of the Wookiees. Its lands are covered in forests, swamps, and towering karst mountain formations that dominate the landscape.

Due to different environmental pressures across the planet, the Wookiee species would have likely diverged over a long period of time.

Ultimately, the most successful Wookiees would end up dominating the environments and potentially interbreeding with, or squeezing out, the less successful clans, just as it happened with hominins on earth. This implies that in the end, hairy Wookiees were probably better at surviving on Kashyyyk than any potentially bald ones.

Fur Enough

Of all the existing primates, we're the only species to have lost their thick coat of hair. We're the abnormal primates. If Wookiees were like primates, their hairiness would be a normal, and possibly expected, attribute. Their hair would also be predominately of the thick terminal hair variety.

However, Wookiees might not be primates. They could be a kind of humanoid dog species; like the Mog from the 1987 Star Wars parody, *Space Balls*. In this case, they might have a fur coat rather than a coat of primate-like hair.

Fur has an undercoat consisting of what's called "down hair." This type of hair is short and mainly used for insulation. The outer visible coat of hair consists of longer guard hairs. They help protect against rain and UV radiation from the Sun. Additionally, there's the medium-lengthed awn hair, which has an intermediate role in insulation and protecting the down hair.

Like primates, dogs can also range from being extremely shaggy-haired like Chewbacca to looking naked like humans.

Although fur can offer a dog added protection from abrasion and insect bites, a lack of it has meant the dogs are generally cleaner with a reduction in odor and parasites. Hairless dogs tend to come from hotter climates. Could this be a factor in the evolution of Wookiees and humans?

Thermal Regulation

In the hot African savanna, a bipedal animal that had less insulating hair could radiate heat away more effectively. In addition to this, as human brains got larger, there was more risk that they could overheat.

Anthropology professor Nina Jablonski has revealed how humans may have lost their fur to help reduce this heat load on the brain. She describes how, among other factors, "shedding our body hair was surely a critical step in becoming brainy."

Sweating helps to cool us down, but if the hair is thick, the excretions from our eccrine sweat glands would quickly matt the hair, hampering heat loss. Therefore, a rise in hairlessness would be of an increasing benefit.

Sweating also carries heat away more effectively in drier environments, so a humid environment would also hamper heat loss through sweating.

The Wookiee home planet Kashyyyk was a forest-covered planet with a temperate climate. It's possible that for most of the year their home planet had relatively low temperatures, depending on whether it had seasons or not. This could favor the need for a thicker coat on Wookiees, just like with woolly mammoths on Earth.

In the heat-regulation model of hair loss, the Wookiees might still have fur because their homeland wasn't very hot and sunny. However, highly exertive activity while working or running would have probably caused them to overheat in all that fur. Dogs can pant to reduce some of their internal heat, but this isn't something we see Chewbacca or other Wookiees doing.

So is it possible that Wookiees have a different-sized brain?

Our brain capacity averages 1,400 centimeters cubed. Our earliest hominin ancestor, *Homo habilis,* had a brain capacity less than half that. They potentially formed the first family units with hunting males and stay-at-home mothers. Additionally, they may have also used simple language. This is similar to the Wookiees.

However, there's no way we could find out whether a *Homo habilis* individual placed in our modern society could succeed in the same way a Wookiee does in Star Wars. If they could, though, then it would also support the case that Wookiees could have similar-sized brains, and therefore have fewer problems with it overheating than us larger-brained creatures. In this way, they may not have needed to lose their extra fur to combat excessive heat.

Ectoparasites and Hair Loss

Mark Pagel and Walter Bodmer presented a theory on ectoparasites and the evolution of hairlessness where it is proposed that fewer parasites could have been more advantageous to humans than actually having hair.

Lethal diseases from blood-sucking insects can lead to premature deaths. A genetic mutation causing less dense fur could increase an animal's chances of survival, causing the hairless mutation to survive and become more prevalent. Over many millennia this could lead to a population of more naked hominids.

They link it with the idea that living in close quarters in "home bases" or lairs around 1.8 million years ago may have increased ectoparasitic activity in humans. Early hominins may have started losing fur by about 1.2 million years ago.

They say their hypothesis can also explain why women are less hairy than men, suggesting that the women were more likely to have spent more time in the home bases and therefore had a higher risk of ectoparasitic infestation. As such, hairless women could have a better survival rate, giving birth to more offspring with the same advantage.

Wookiees are a communal species that build their homes in the gigantic Wroshyr trees, which cover the Kashyyyk landscape. A downside to their communal living is their close contact, which can more readily spread the disease-carrying parasites. Pagel suggests that "we would expect them to have the ectoparasite levels of other big ape-like species such as chimpanzees or gorillas, which also live in social groups."

Like chimps and gorillas, Wookiees indulge in grooming, which to them is also apparently the highest way to compliment. With the thickness of Wookiee hair, we can imagine they would have to spend a great deal of time grooming, unless they had a natural repellant or some kind of technology that could function similar to a flea and tick collar.

Han and Chewie's relationship may have been a lot different if the *Millennium Falcon* were constantly infested with Wookiee-borne parasites.

To summarize: Regardless of whether Wookiees have hair or fur, a hot environment would be the main driver towards hairlessness. The abundance of forests and lack of deserts implies that, unlike the African savannah, Kashyyyk is not a hot and dry place. This would make sweating less efficient. If this is combined with a relatively cool climate most of the time, then insulating fur may have been more advantageous to the Wookiees than losing it to improve sweating. They may also have smaller brains, which could lessen the need for improved heat loss via bare-skinned sweating. However, they would still have to deal with ectoparasites. If this is a strong instigator of hairlessness, they are still hairy because they have effective ways to dispose of the ectoparasites.

HOW DOES WATTO MANAGE TO STAY ALOFT WITH THOSE FUNNY LITTLE WINGS?

Greek mythology had the flying horse, Pegasus. Now Star Wars has given us a modern mythological version in the form of Watto, the Toydarian junk dealer.

We first meet Watto in Mos Espa on the desert planet of Tatooine. Qui-Gon Jinn is looking for a T-14 Hyperdrive generator and Watto is the flying, talking, alien scrap dealer who can provide it for him.

Now you may be familiar with the old adynaton "When pigs fly." Like every adynaton, this phrase is meant to point out the implausibility of something happening by comparing it to an even more outlandish idea.

In this case, Watto is pretty much the pig and he *is* flying, despite his unwieldy appearance. Let's take a closer look at what's needed to get this pig to fly.

Winging It

He's got the mouth and belly of a boar, the snout of a tapir, and ludicrously small wings attached near his shoulder blades. Despite the size and positioning of his wings, he flaps them at a rate that is visible to the unaided eye. About four or five flaps per second. This seems sufficient to make him hover.

It doesn't take an expert to see there might be a problem here.

However, an often-repeated folklore claims that bumblebees shouldn't be able to fly according to science, yet they do. This was never a real scientific claim, though. The first known appearance of such an idea has been credited to a French zoologist and aeronautical engineer called Antoine Magnan, who wrote:

> *"First prompted by what is done in aviation, I applied the laws of air resistance to insects, and I arrived, with Mr. Sainte-Laguë, at this conclusion that their flight is impossible."*

Despite their calculations, a stroll through a park in spring can rest us all assured that the bumblebee and many other insects can indeed fly. Just because we know how to make planes fly doesn't mean the rules can be applied in the same way to describe flight in nature.

As such, when looking at the possibility of Watto's flight, we should take our cues from nature as well as theory.

Taking to the Skies

To fly, Watto needs to overcome the gravitational attraction that causes him to experience weight. For this, he needs to create an opposing force, in this case, an upward force.

When it comes to flight, this force is referred to as lift and the balance between lift and weight forces will affect whether an object goes up, down, or just hovers.

The following equation gives an idea of some of the major factors involved in generating a lift force.

$$\text{Lift} = Cl \bullet \tfrac{1}{2} \bullet \rho \bullet V^2 \bullet S$$

Don't worry if it looks complicated; it's just saying that lift (on the left) depends on the factors on the right. These are density of the air (ρ), the speed of the wing through the air (V), and the area of the wing (S).

The "Cl" is a number that relates the shape of the wing to all of the other factors; it's called the coefficient of lift. All you need to know for the sake of this argument is that if it's possible to make any of the factors on the right bigger, you get more lift.

So in order to fly, all animals (including Toydarians) have to develop a means of generating and controlling lift forces. The eventual nature of their flight will be down to a playoff between the factors needed to create lift, i.e., the variables on the right-hand side of the equation.

Nature's Fliers

The debate is still on as to how animals first developed the ability to fly, but it is generally achieved through two methods: gliding or powered flight.

Gliding can be seen in a variety of species including flying fish, flying frogs, and flying squirrels. However, their flight is limited and pretty much a form of guided falling. For an animal to gain any real altitude or hover in the air like Watto, then it needs to master powered flight.

Life has stumbled across the ability to create lift through powered flight on a number of occasions and it always involves wings. The wings are flapped to achieve lift by increasing the speed (V) of the air over the wing.

Wings first evolved around 350 million years ago on insects, which are invertebrates. Wings would later independently appear on three different groups of vertebrates.

Around 225 million years ago the pterosaurs appeared, followed many millions of years later by birds. Lastly, flight in mammals came with the appearance of bats. The internal skeleton of vertebrates is generally heavier than the exoskeleton on insects, and the wings and flight muscles are also arranged differently. So how does Watto compare?

Watto Versus Nature

Watto's wings are attached to his back like an insect. However, insects are hexapods, meaning that they have six legs. They also tend to be lightweight with exoskeletons. Instead of lungs, they breathe through holes along their abdomens called trachea. In order to get enough oxygen circulated through their bodies, insects can only grow to be a limited size.

Currently, the largest known insects that ever existed were the now-extinct Meganisoptera. These were dragonfly-like insects that had a wingspan of more than seventy centimeters and are estimated to have weighed almost half a kilo.

There are so-called insectoid species in Star Wars such as the Geonosians. They're six feet tall with exoskeletons; however, with only four limbs, they are tetrapods—meaning four-footed. This means they aren't technically insects by definition, which are always six-footed.

Toydarians are also a four-footed species, and having belly buttons puts them more into the realm of flying mammals, like bats. The positioning of their wings is different from bats, though.

Bats and all other known flying vertebrate species use their upper limbs for their wings. Toydarians do not. Their upper body limbs only function as arms while their wings are connected dorsally, similar to insects. They are a mish-mash of structures evolved for flight in both invertebrates and vertebrates, which creates a problem in determining what category they could fit into.

Insects have rigid, delicate wings, powered by muscles connected to the inside of their exoskeletons. Watto, on the other hand, has wings that are not delicate or rigid at all—appearing more like elongated, floppy elephant ears until he engages them for flight. He also appears to be supported by an internal skeleton, the defining feature of vertebrates.

To operate his wings, Watto would need powerful muscles attached to the base of his wings. This is how vertebrates power their flight. Flying vertebrates tend to have large breast muscles attached to a keeled sternum (breast bone). As Watto's wings are between his shoulder blades, he would need similar-sized muscles to be attached there instead. This would pose potential problems to his ability to use his shoulders and arms while flying.

Air Speed and Flapping

The heavier the animal, the more its wing loading, i.e., the amount of lift force it needs to generate based on its wing size. This is because as an animal gets larger, its mass increases faster than its wing area.

To compensate for the increased weight the animal will have to move through the air faster, or else flap its wings faster if it's moving at a slower speed. Both of these methods will increase the velocity (V) in the lift equation.

As large wings require more energy to flap faster, it's far more economical for larger animals to fly fast rather than flap excessively. This is why they tend to be seen soaring, not hovering.

Now, if Watto is as heavy as he looks, he would need to beat his wings extremely fast to attain his hovering ability, which would put massive strain on his muscles, joints, and wings. Nature adheres to this limitation with light flyers, such as mosquitoes, flapping with a frequency of more than six hundred beats per second, while the much larger Andean Condor flaps just one to three times a second.

Watto can be seen flapping his wings at a rate of about four or five flaps per second. Although this is consistent with a largish animal moving through the air at speed, it just isn't fast enough to support hovering. Hummingbirds flap around fifty times a second to hover, while bats can manage hovering at around ten flaps per second.

After examining all of this information, it seems fair to say that Watto —and his fellow Toydarians—are unlikely flyers that seem to disregard anything nature has to say about flight in animals. It's quite possible that we'd have a better chance of getting a pig to fly.

However, something may have been overlooked: he could have a better chance at flight on a planet with a lower gravity and a denser atmosphere. A lower gravity would reduce the amount of lift needed to achieve flight, while a denser atmosphere could enable him to generate more lift with each flap (according to the flight equation). A candidate location would be Saturn's largest moon, Titan, which has both of these features. Although, he'd need an oxygen mask!

PART IV
TECH

HOW LONG BEFORE WE GET A STAR WARS SPEEDER OFF THE GROUND?

Admit it. You've imagined yourself on a Star Wars speeder.

Perhaps, in your daydream, you fly through the dense forests of Endor. Or maybe, like Rey, you glide over the desert sands of Jakku in a scratch-built mammoth of a speeder.

Speeders in Star Wars look like rocket-powered hover-scooters, which use engines that reach speeds up to 310 miles per hour. Speeder pilots need to be skilled, as the crafts achieve high velocity and maneuverability at the expense of protection for their riders.

But how do speeders work?

If they hovered like hovercraft on Earth, then their operation would be relatively simple. They'd need an oval (or roughly rectangular) platform, a motorized fan, and a large skirt of material to trap air underneath the craft. The air cushion (or plenum chamber) underneath the hovercraft forms a ring of air, which circulates around the base of the skirt. This then insulates the air cushion from the lower pressure outside the skirt. In other words, the ring of air keeps the air under the craft from escaping.

Once buoyant, all you need is propulsion and steerage.

Luke's X-34 Landspeeder had steerage and propulsion in the form of the three air-cooled thrust turbines, which sit horizontally at the back of his speeder. But there is certainly no skirt in sight. Another example would be 614-AvA speeder, the Lothal Speeder Bike used by the Imperial military. Or the 74-Z speeder bikes used in the Battle of Endor. They seem to be simply composed of a sleek-looking seat and steerage shaft, along with tail fins and stirrups. Not much hovering going on there.

And magnetic levitation (or maglev) is unlikely to be the solution either.

If you've ever played with magnets, you know that opposite magnetic poles repel. That's the basic idea behind magnetic levitation. Using powerful electromagnets, a maglev train, for example, floats upon a magnetic field that sits between train and track. But there's no track in the woods of Endor, and no rail in sight on the sands of Jakku.

Remember that fleeting scene where Han Solo's frozen form floats through the corridors of Cloud City? Ain't *that* as cool as carbonite? Perhaps this scene is key to the mystery of speeder power. You can consider the physics of flight, uplift, and thrust. You can even mull over maglev. But there's something special going on in the science of Star Wars. When Luke parks his landspeeder and switches off its engine, doesn't it continue to float just like the carbonited Han?

The Star Wars secret is repulsorlift technology.

Repulsorlift was a technology that enabled a craft to hover, or even fly, above the surface of a planet by pushing against gravity, producing thrust. Many have taken this repulsorlift engine design to conclude that speeders use an antigravity device.

Antigravity

Antigravity is one of the great science fiction dreams.

The idea of a force that opposes gravity emerged in the late 1800s. Typically, writers imagined devices allowing people—or objects—to hover or to be boosted about. An antigravity principle, known as "apergy," was used to send spaceships to Mars in some early tales. Less romantically, in another early story, an antigravity ointment is smeared on the hero's space vehicle.

Unsurprisingly, we owe the most famous antigravity device to H. G. Wells. *First Men in the Moon* ingeniously describes how antigravity shutters made of "Cavorite," a metal that shields against gravity, is used to send rockets to the Moon. Buck Rogers even had an antigravity belt.

When we take a closer look, antigravity is rife in Star Wars. Not only with Luke's "hover" speeder on Tatooine, and the sleek speeders on Endor, but also with Jabba's *Khetanna*, the massive sail barge, with a crew of 26 and a capacity of 500 passengers. Also, let's not forget the frozen, floating Han or the Single Trooper Aerial Platform (STAP) vehicles piloted by single battle droids.

From bulky to bijou in Star Wars craft, repulsorlift technology is king.

Repulsorlifts

So we assume that, like the science fiction stories of old, Star Wars craft use some kind of anti-gravitational field to float in the air.

It looks effortless, too. Just picture Luke's parked speeder, Solo in stasis, or every single craft lined up for the start of the pod race in *The Phantom Menace*. They float without artifice, easily resisting the planet's gravitational field. To stay at a certain height above a planet's surface in that way the speeder, or any type craft, would need to apply an equal and opposite push to the gravitational force of the planet (the idea of balanced forces is one from old British scientist, Isaac Newton, of course). For acceleration in an upward direction on the planet, the craft would need to apply an even greater force (another concept of Newton's).

Since, according to Einstein, gravity is just curved space, all that's needed to create antigravity is to simply bend space the other way. If the geometry of space can be bent to your will, then you have antigravity and the ability not only to float freely, but also to tear across the sky.

Tricky, though.

Whereas all mass makes gravity, it's not easy to come by material that makes antigravity. But we know of a material which *may* be a willing candidate: exotic matter. In theory, exotic matter has negative energy, or negative mass (matter whose mass is of opposite sign to the mass of normal matter, for example, −2 kilograms or −4 pounds). Thus, exotic matter should create the reverse effect of gravity, and it could be used to cancel out the weight of a speeder, or any repulsorlift craft.

Let's Build a Speeder

Imagine you're some kind of engineering wizard, like the wunderkind Anakin Skywalker. You've mastered pod racing, and your next pet project is your own personal speeder. You've settled on your sleek and retro design, something straight out of *Forbidden Planet*, or the equivalent of 1950s movie sci-fi in the Star Wars realm. You know the kind of thing: all retro-futurism, Perspex windshields and go-faster tail fins.

However, the crucial bit of the build is the repulsorlift. Just how much exotic matter do you add to the mix? You simply measure the mass of your craft, and pop into the speeder an equal mass of the exotic stuff. Bingo, the resultant mass of your speeder is zippo. Now, with no effective mass, your speeder won't be pulled down by the planet, or repulsed from it. When you park it, just like Luke, your speeder will rest at whatever height you left it. With thrusters on board, too, you're now free to roam the galaxy!

Imagine more ambitious and complex builds, such as Sebulba's huge pod racer, Jabba's *Khetanna,* or even the repulsorpods, the hovering balconies for members of the Galactic Senate. To offset any additional mass in each case (including passengers), you'd simply need to add the corresponding amount of exotic matter. This would prevent *Khetanna* from sinking into the sand, or the delegates descending rapidly into the debating chamber (we wouldn't want the politicians to fall from grace now, would we?)

One last remaining technical detail remains: What exactly *is* exotic matter? Trouble is, no one really knows. A broad definition of exotic matter is "any kind of non-baryonic matter." Normal matter is made of baryons, subatomic particles such as protons and neutrons. Exotic matter is simply made of different stuff. We're just not sure what that stuff might be. Yet.

Take-Off?

Nonetheless, some version of Star Wars speeders may soon be upon us.

A number of business companies are currently trying to create working versions of hoverbikes. The *Aero-X* hoverbike was created by Aerofex (an aerospace start-up company based in Los Angeles) and is designed to carry up to two people. Set for imminent release, the vehicle rises ten feet from the ground and can travel up to forty-five miles per hour. It's expected to weigh in at 785 pounds and be fifteen feet in length, running for around seventy-five minutes on a full tank of fuel, plenty enough for short trips over the sands of Jakku or through the woodlands of Endor.

The hoverbike has two horizontal wheels which act as ducted rotors. They are made of carbon fiber blades, which operate in a similar manner to the open rotor of a helicopter, but with much tighter control. The *Aero-X* doesn't fly with the same energy efficiency as a helicopter, due to its rotor blades being shorter, but it's much smaller in size and safer near humans. The hoverbike doesn't produce the 'brownout' of helicopters, as it's designed to be able to operate near people without blowing any significant amount of dust.

If Darth Maul is on your tail and you have a need for speed, UK-based Malloy Aeronautics' *Hoverbike* claims to reach speeds of more than 170 miles per hour at the same altitude as a helicopter.

Both Aerofex and Malloy Aeronautics' hoverbikes use ordinary gasoline, but environmentally conscious Star Wars aficionados may soon have futuristic travel options, too. Bay Zoltan Nonprofit Ltd., a Hungarian state-owned applied research institute, has created an electric battery-powered tricopter called the *Flike*.

All three vehicles—*Aero-X*, *Hoverbike*, and *Flike*—are firmly in the design stage.

STRUGGLING WITH STORMTROOPER UNIFORMS IN BATTLE: IS THAT WHY THEY'RE SUCH TERRIBLE SHOTS?

The Star Wars franchise is often winning accolades for ingenious costume design.

In these days of CGI dominance, you can do almost anything with makeup and costume. So what exactly makes the Star Wars wardrobe department special, critics ask? They sing about the cavalry/cowboy fusion in the style of Han Solo. They salivate over C-3PO, a robot design that takes its inspiration from Maria, the Maschinenmensch in Fritz Lang's 1927 sci-fi movie classic, *Metropolis*. They remain in rapture for Lucas' *pièce de résistance* in Darth Vader: an evil all the more terrifying and unknowable due to the anonymity and mystique of his faceless helmet.

In spite of all of the aforementioned costume design feats, the images that accompany the articles and the accolades are often of the Star Wars stormtroopers.

American fashion and lifestyle magazine, *Vogue*, even named the stormtrooper suits in their top ten Star Wars catwalk crossovers. Witness pictures from São Paulo Fashion Week, when Brazilian label *Triton* updated the classic stormtrooper look for its fall show in 2015.

But has this coterie of fashion aficionados, film critics, and journalists actually tried to *wear* the stormtrooper suit? It's high time for an exposé. We need to lift the lid on what life might be like inside one of those suits, with all its attendant challenges at work, rest, and play.

Bedecked Like a Buckethead

First up, what are the stormtroopers actually wearing in the films? The official stormtrooper armor, the standard armor worn by the Imperial trooper, is a white plastoid composite worn over a black body glove (similar to the fit of spandex, but much tougher). The armor represents some of the best in the Empire, and is dreaded by rebel freedom fighters.

The suit consists of eighteen individual, overlapping plastoid plates and synthetic leather boots, which aid mobility. A reinforced alloy plate ridge supports the user's upper thigh (though exactly how, and in what way, isn't stated in the "manufacturer's guidelines"). A sniper position knee protector plate over a wearer's left knee help to improve precision when crouching, which assumes all stormtroopers are right-handed shots. However, it's safe to presume the sniper plate is switchable to the right knee. The suit also helps "disperse energy," and protects the wearer from glancing blaster bolts. Note the word "glancing" for later.

Species other than human do not escape the delights of the stormtrooper suit, either. While the great majority of manufactured armor is fitted for humans, other forms are made to suit other body types. The thought of Yoda suited up in such a way is quite amusing. And surely, the sheer mass of Jabba the Hutt would challenge any composite plastoid to breaking point.

For all species, a soft click informs the wearer of whether or not they have correctly attached the armor. That's some small comfort, at least. Also reassuring is the manufacturer's promise that the armor is impervious to projectile weapons and blast shrapnel. (That's some pretty tough composite plastoid, right there.) However, the manufacturers do admit that the plating made running difficult, and is vulnerable to direct hits from Cyclers (crude yet reliable rifles that fired solid projectiles) and blasters, whenever the hit is not "glancing."

Disappointingly, the manufacturer makes no mention of the problematic crotch plate, which could prove disastrous in times of urgent need to use the bathroom, or the random and spontaneous prospect of romance, especially since there's also the black body glove to contend with.

Buckethead in the Field

Once out in the field, the rank-and-file stormtrooper can be confident in his kit.

Remarkably, the armor is able to protect its recruit in most extreme environments. This included the forests of Endor, the deserts of Tatooine, the icy wastelands of Hoth, and even a limited exposure to the vacuum of space, just in case a recruit finds himself adrift, or cascading *from* the Death Star and *down into* the forests of the Sanctuary Moon. It's not clear whether the suit would survive an impact on the satellite from such a height. That would surely depend on the Jabba bench test.

The secret of the suit's variability lies in its midriff—the armor's torso plating included environmental controls on its midsection. The underlay of the black body glove is vacuum-sealed, and manufactured from a smart material that adjusts not only to the recruit's body heat, but also to his external temperature.

The stormtrooper suit isn't a million miles away from the Extravehicular Mobility Unit. The EMU suit was used by crewmembers on NASA's Space Shuttle, and by members of the International Space Station. The EMU protects the astronauts from the dangers of space and worlds other than Earth with a suit that has approximately fifteen different layers, including smart materials, such as spandex, Gortex, and Kevlar. The EMU protects the astronaut's body from contact with space junk, micrometeoroids, and radiation. Like the stormtrooper suit, the EMU was created in different segments that fit together.

There, the similarities seem to end. As the EMU not only allows free movement and maximum comfort, something sorely lacking in the stamp of an average stormtrooper suit, it also boasts a Maximum Absorbancy Garment (MAG), which collects the necessary fluids when nature calls.

Perhaps the lack of a MAG equivalent in the stormtrooper suit explains the often-erratic performance of Imperial recruits.

Buckethead? Are You Receiving Me?

NASA's EMU is something of a steal at twelve million dollars apiece. What price would a stormtrooper suit with all its add-ons and extras be in American currency?

A hefty one is an understatement. Here's why: First, it includes a reinforced combat helmet, which not only features an integrated comlink, audio pick-up, two artificial air-supply hoses, and a broadband comms antenna powered by a single cell, the helmet also has on-board filtration systems for extracting breathable atmosphere from polluted planetary environments. The helmet's visual processor helps the recruit see in darkness, glare, and smoke, though it limits the wearer's field of vision. This may go some way to explain why many troopers are such mediocre marksmen. It does not seem to help that when firing a blaster, the helmet's visor polarizes against glare.

Adding to the complexity of the helmet, a built-in heads-up display also provides targeting diagnostics, power levels, and environmental readings at the corner of the wearer's eyesight. A recruit can also access data on various military subjects and civilian organizations on the helmets display. The interior of this combat helmet, an Imperial hybrid of Google Glass and Oculus Rift, sounds somewhat akin to the seeming chaos at air traffic control. An average buckethead may not know whether he's coming, going, or being seduced by the dark side.

Perhaps aware of the cacophonous interior of the combat helmet, Imperial Command actively discourages nonessential chatter, which is strictly off-limits while on-duty. Stormtrooper helmets record all that is said by its user, sending it to Command for review after downloading from the armor's data banks. It is sheer speculation, and mildly apocryphal, to suggest that those at Imperial Command may spend many a happy hour watching blooper reels of stormtroopers in combat.

Surely Not, Buckethead?

A couple more interesting footnotes may go some way to explaining the enigma that is stormtrooper performance in the field.

We've already exposed the manufacturer's admission that the plastoid plating makes running difficult, and that limited visibility in the combat helmet undoubtedly hinders precision shooting.

Two further possibilities emerge:

First, that suit must stink. Stormtroopers are expected to remain in uniform at all times. This command is considered crucial for keeping Imperial subjects under the cosh, and representing the might of the Empire at all times. By wearing one's armor, one represents the Empire, not the individual. But the stench of all those suits must surely make the prospect of fleeing one's comrades, those massed ranks of reeking recruits, a more favorable course of action than staying put in the heat of battle.

Second, the armor's utility belts are equipped with a variety of features, which included macrobinoculars, a grappling hook, and a thermal detonator. But check this out: the controls to the detonator are not labeled, to prevent enemy troops from activating them. Surely this is a recipe for incompetence, if not complete disaster, on those frequent occasions when recruits are called into battle. After all, are these not the elite shock troops of the Imperial Army?

HOW COULD DROIDS LIKE BB-8 HELP US EXPLORE MARS?

It all started with the first trailer for *Episode VII.*

The camera pans over dunes of sweeping sand. Finn suddenly jerks his head into frame. Then an astromech droid, moving incredibly swiftly—with a drive system that sits miraculously static with respect to the rolling spherical body—tears across the sands of Jakku.

A rolling ball robot with a floating head! Such a purely appealing prospect quickly made BB-8 the unofficial droid mascot of *The Force Awakens.*

But what if we swap the sands of Jakku for the sands of Mars? How might this orange-and-cream droid help us explore the Red Planet?

Astromech Spec

Astromech droids are a class of repair droid that are used as automated mechanics on starships. The droids are compact, and have a small "operational footprint," most being approximately one meter tall. They are tooled-up through special limbs that are hidden in recessed compartments on the droid body.

Many starfighter spacecraft rely on astromech droids to act as co-pilots. As the droids sit in astromech sockets, exposed to space, they control flight and power distribution systems, as well as calculate hyperspace jumps and perform routine maintenance.

Chatting astromechs are nothing like prattling protocol droids. Astromechs communicate only in writing via computer systems or through binary—that familiar sequence code of clicks, bleeps, and burps. The BB unit is a new class of astromech created some time after the Battle of Endor, with the main new innovation being its spherical body.

What kind of jobs could a BB unit do on Mars? They potentially include: planetary habitability studies preparing for future human exploration, investigating Martian climate and geology, and confirming whether Mars has ever offered environmental conditions favorable for microbial life.

Journey to Mars

NASA is on an ambitious journey to the Red Planet, which will include sending humans in the near future. Current and future robotic spacecraft are leading the way, and will prepare foundations in advance for those human missions.

One recent robot innovation is the use of "relay units." In a job that seems perfectly suited for a BB unit, the relay unit transmits radio data from the Martian surface to an orbiter passing overhead. Relay of information from Mars-surface craft to Mars orbiters, then from Mars orbit to Earth, helps sample much more data from the surface missions than was previously possible. You just have to remember there's a time lag in data coming from Mars. This can be anything between three and twenty-one minutes, as the Earth-Mars distance varies from about 55 million to 378 million kilometers. Put more simply, sometimes Mars is on the same side of the Sun as us, sometimes it's not.

A BB unit could also scour Mars for radiation potentially harmful to humans, if they were to be exposed to it.

There's an armada of rovers and robots already on, and orbiting, Mars. They have helped dramatically improve our knowledge of the Red Planet. A BB unit could help pave the way for future human explorers by getting better measurements of radiation data from the Martian surface. The data would help plan how to protect the astronauts when they get there.

Journey to the Center of Mars

Perhaps a BB unit's coolest Martian mission would be caving.

In this book's entry on the links between Hoth and Mars, we cited NASA's stunning discovery in 2007 of a series of tell-tale "skylights" in the Martian surface. The tale they told was of subterranean tunnels under Mars, some potentially cavernous. These tunnels, made by the journey of ancient lava under the Red Planet's surface, may be the secret to understanding Martian history.

Digging down is the key to divining Mars's past. Geology works in layers. The more you dig down, the further into the past you delve. It's a little like the famous story *Journey to the Center of the Earth*, by the nineteenth century French sci-fi writer Jules Verne. In the book, a professor gains entry to the Earth's interior through the cone of a defunct volcano in Iceland. As the professor delves into the Earth, he examines the rocks, and his expedition also becomes a quest into the depths of evolutionary time.

BB-8 could be sent on a similar mission.

The skylights are doorways into Martian sinkholes. Since geology works in layers, the Martian sinkholes are like the volcano cone in *Journey to the Center of the Earth*. The sinkholes may expose hundreds of feet of Martian strata. So, if a unit like BB-8 were to explore the tunnels, the story in the stones could be deciphered without jeopardizing human life.

The sinkhole spelunking would save untold amounts of time. That's because previous plans for understanding Mars's past involved reading the rocks, little by little, layer by layer. Previous landing sites have included ancient Martian impact basins, rising out of crater floors. The data would have to be read from miles-high mounds of layered rock. But an even richer seam to data mine would be hidden underground.

BB-8 Goes Spelunking

How would BB-8 get into the tunnels?

One way would be down through the sinkhole itself. Although that might be a little tricky as BB-8 would most likely need some kind of repulsorlift tech to prevent it from plummeting down into the Martian depths. A second option would be drilling, or even blasting, through the roof of the lava tube. BB-8 would need some kind of diamond-tipped and heavy-duty drilling limb, hidden in one of those astromech compartments.

But, for argument's sake, let's hypothesize the eco-option. No point in messing up Mars.

BB-8 could simply enter the tunnels through one of the rille entrances, which are terminal openings to uncollapsed sections of lava tube networks. Once in the cosmic catacombs, BB-8 could begin work. Here in the tunnels are potential habitats, which are more shielded from the radiation that constantly blasts the Red Planet's surface. The subterranean tubes are also better protected from meteorite impacts, and would have more stable temperatures through the Martian day-night cycles. These conditions make the lava tubes good not just for preserving data samples, but also good for human habitation.

Life in the Universe

The catacomb habitat on Mars won't just act as a human colony in the near future.

It might also have hidden in its depths the answer to one of the fundamental mysteries of the cosmos: Does life exist beyond the Earth? So BB-8 can do some serious science looking for clues that might uncover evidence of life. BB-8 would be programmed with the knowledge that water is the factory of life, giving chemicals the right kind of environment to combine and thrive. The elements hydrogen and oxygen make up water. Separate, they are explosive. But together they combine to make the safest of materials, one that is unchanging over a wide range of temperatures. That's why BB-8 would "follow the water" when looking for life on Mars.

However, water is not the only condition for life. Life must also be shielded from a hostile environment. In the past, Mars had a magnetic field, which protected the planet from cosmic and solar rays. BB-8 would be tuned to search for microbial life, hiding deep under the crust in the cracks and caves of the catacombs, sheltered from today's much harsher conditions on the Red Planet.

Caves, Not Canals

In the late 1800s, the idea of alien life was a very popular one. The stage for the drama and scandal that unfolded was not, of course, the fictional sands of Jakku but the factual sands of Mars.

The idea of life on Mars was so popular that many people got completely carried away. Some scientists interpreted marks on the Martian surface to mean that there were *actual* Martians. The man at the center of this Mars scandal was Italian astronomer, Giovanni Schiaparelli. The problem began in 1877, one hundred years before *A New Hope*. Schiaparelli had observed Mars on a number of occasions previously, and had described seeing what he called "seas" and "continents." But the trouble started one night when, at the telescope, Schiaparelli noted long linear features on the Martian surface, which he called *canali*, meaning "channels."

As has now become legend, a willing Boston entrepreneur, Percival Lowell, became convinced that Schiaparelli had identified artifacts of a dark alien race. In several publications early into the twentieth century, Lowell described the "canals" as clear evidence of a sophisticated civilization. Martians were using the canals, Lowell insisted, to transport water from the polar caps of Mars down to the parched equatorial regions.

Curiously, the Mars scandal has come full circle.

There may, after all, be life on Mars, although admittedly microbial. In the near future, however, human life is predicted to make its way there. The lifeblood of the Red Planet may not be canals, but the "channels" of the lava tubes and cave systems may make all the difference to human habitation.

NASA and other space agencies are developing the capabilities needed to send humans to Mars. Scientists all over the globe are working hard to develop the technologies that will one day be used to live and work on the Red Planet, and safely return home. We may not yet have BB-8 units to help pave the way, but we know what science needs to be done for the next giant leap for humanity.

IS THE EMPIRE WATCHING YOU?

The Dark Lord of the Sith marches into the blinding light of a corridor of the Rebel blockade-runner.

Face obscured by a grotesque breath mask, and flowing all-black *Schutzstaffel* robes, the sight of the Dark Lord jars starkly against the faceless, white-armored fascistic suits of the Imperial *Sturmabteilung.* Rebel rivals back away from Vader, as a deathly quiet sweeps through their troops. The Dark Lord is the personification of power. No one dares catch his "eye." There seems to be no freedom to think, never mind act. And no hope for the future.

The very tools of totalitarianism are writ large in Star Wars. A Galactic Empire wielding rigid control of its subjects. Uniforms that evoke dark shadows of Europe's fascist past: the jodhpurs, the helmets, the jackboots. The use of excessive force and violence for minimal aims. The atavistic fear. A dissolved democracy, now a dictatorship, led by an all-powerful ruler.

But the politics and technology of control can also be subtle.

Consider the dark side of the Force. Those who ply the dark side draw power from primal emotions, such as fear, anger, and hatred. Darth Sidious says that the source of his dark power is "the universe beyond the edges of our maps." Unseen. Undetected. The power of the dark side is penetrating and pernicious. Even a Jedi raised from birth in the traditions of the Jedi Order could quickly become corrupted by the lure of the dark side and its seductions.

In both brutal and subtle varieties, Star Wars is a timely reminder that totalitarianism is a past, and present, prospect.

The Sith, the Death Star, and the Force

You would hope that we'd recognize such totalitarianism when we saw it. Film and fiction have given us plenty of warning, the most famous example being George Orwell's classic book, *1984*. There's something very Orwellian about the Empire in Star Wars, although the flavor is upbeat science fiction, of course, and far from the downbeat dystopia of *1984*. Both works hugely captured the public imagination. Both incorporate the theme of big government gone mad with lust for power. And each fictional regime has its champion of evil—Big Brother in *1984*, Darth Vader in Star Wars.

It was Orwell's fiction that famously foresaw the kind of totalitarian control meted out by the Galactic Empire. The insidious nature of *1984*'s culture of surveillance stems from its telescreens and Thought Police. In Orwell's marvelous words: "The Beehive State is upon us, the individual will be stamped out of existence; the future is with the holiday camp, the doodlebug and the secret police."

Perhaps Darth Sidious would have morphed Orwell's words into something like: "The Galactic Empire is upon you, the individual will be stamped out of existence; the future is with the Sith, the Death Star, and the dark side of the Force."

Indeed, Darth Sidious's rise to power is a classic lesson in slow, creeping totalitarianism. His role in transforming a democracy into a dictatorship, consolidating power, and going from Chancellor to Emperor, has a Roman feel about it, as well as allusions to Napoleon and Hitler. Yet, George Lucas has also suggested that the Galactic Empire was largely derived from America during the time of the Vietnam War. More specifically, the period of American politics when Richard Nixon was President, as 1973 was when Lucas first started working on *A New Hope*.

As Palpatine, Sidious legitimizes authoritarian rule within the Galactic Senate by saying that corruption is hampering the powers of the head of state. It's a common tactic used by contemporary politicians. The allusions to external threats are simply self-serving; they allow the super-State to maintain hysteria within its borders.

The Technology of Control

Then there's the mastery of the machine.

In *1984*, science's machine mastery is so complete that utopia is possible. However, poverty and inequality are maintained as a means of sadistic control. The visual medium of monitoring in the two-way telescreen is a brilliant evocation of the all-seeing eye. In Orwell's book, it is politicized into a technological nightmare. As citizens dutifully follow the daily exercises on the telescreen, they are at the same time observed by it.

In Star Wars, any physical machinery of control is obsolete, tossed into the trashcan of history by the dark side of the Force.

Through the use of the dark side, the Sith are able to invoke that ever-present aspect of totalitarianism—there is no privacy, and you are never alone, ever. There is no need for the telescreen, or the Thought Police. The Sith *are* the Thought Police, and they can use the penetration of the dark side to divine intent. "You claim you are innocent. And yet, I can read your thoughts. They testify to your treason, your plans against the state. The Force gives us, with absolute certainty, proof of your disloyalty against the Galactic Empire."

But it's far from personal. With every example of a Galactic subject interrogated and suppressed by the dark side of the Force, the whole time everyone else gets the message: *You* could be next. So be a good citizen of the Empire, mind your own business, but also keep a beady eye on others lest you become embroiled in their treason. It's like a web of fear.

Web of Fear

So the science of the dark side is the ultimate in state surveillance.

And yet, there are other ways in which the science of information can be used to exert political control. Since you began reading this entry, an agency has selected over one hundred terabytes of data for review. That's about 25,500 two-hour HD movies. The National Security Agency (NSA) intercepts telephone and Internet communications of over a billion people worldwide. Their mission is to gather data—on external threats; on foreign politics; on economics and commercial secrets. Reports revealed how penetrating their mission has become. Between 2006 to 2009, 17,835 phone lines were placed on an improperly permitted alert list, in breach of compliance, which marked down these lines for daily monitoring. Only 11% of these monitored phone lines met the NSA's legal standard for reasonable suspicion. Not even Orwell could have dreamt of surveillance this draconian.

According to a report in *The Washington Post* in July 2014, 90% of those placed under agency surveillance in the United States are ordinary citizens, and not intended targets. The *Post* said it had examined documents including emails, text messages, and online accounts that support the claim.

Elsewhere, there are the same patterns. In the United Kingdom, government surveillance of work, travel, and telecommunications is rabid. Actions of its citizens are becoming increasingly monitored through the use of credit card and mobile phone information, as well as through closed-circuit television (CCTV). The first CCTV system was installed by the Nazis at Test Stand VII in Peenemünde Germany in 1942, for observing the launch of V-2 rockets. It has been estimated that there are up to 4.2 million CCTV cameras in the UK, around one for every fourteen people. In the UK, too, we find aspects of Orwell's world of political spin and euphemism. You can also imagine Darth Sidious using such phrases: "war is conflict," civilian casualties are described as "collateral damage," firing employees has become "right-sizing," a fix for a software bug a "reliability enhancement."

Security and intelligence agencies worldwide are involved in the penetration of global surveillance, including those in Australia, Britain, Canada, Denmark, France, Germany, Italy, the Netherlands, Norway, Spain, Switzerland, Singapore as well as Israel, which receives raw, unfiltered data of citizens that are spooked by the NSA.

The Agency

The NSA tracks the global position of hundreds of millions of cell phones, every day.

The tracking enables the mapping of people's movements and relationships, with considerable precision. The NSA also has access to communications made via Google, Microsoft, Facebook, Yahoo, YouTube, AOL, Skype, Apple, and others. Such penetration allows the NSA to collect hundreds of millions of contact lists from personal email and instant messaging accounts each year. Through collaboration or coercion, or else infiltration of numerous technology companies, the agency has also weakened much of the encryption used on the Internet, so that the majority of Internet privacy is now vulnerable to attack.

However, if the NSA is so all-powerful and penetrating, why allow the information to leak? Not ideal, admittedly. And yet the very leaking serves a similar purpose to the tactics of the Empire. Now, everyone knows they are being watched. *They* could be next. So be a good citizen, and knuckle under. Their dark power is beyond the edges of our maps. Unseen. Undetected. Their Beehive State is upon us. The individual will be stamped out of existence. The future is with reality TV, the drone, and the NSA.

COULD A SINGLE BLAST FROM THE DEATH STAR DESTROY THE EARTH?

To answer this question, a Death Star Commander such as Grand Moff Tarkin would need to consider what kind of wicked weaponry his Death Star wielded.

According to canon, the Death Star's main armament was its superlaser—a weapon powered by hypermatter. The use of "hyper" is a common sci-fi trick. Not only does the word hyper *sound* scientific, but it also means "over and above," or "beyond," implying this is no ordinary matter, but some kind of exotic stuff that actual science hasn't yet encountered.

Even so, the rest of the Death Star weapon dynamics are easily understood. With sufficient firepower to destroy an entire planet, the superlaser was only fired once, and only on a fixed target, such as Alderaan. Then the Death Star needed a twenty-four-hour recharge time before it could do any more damage.

Sci-Fi Invented the Atom Bomb

The idea of an ultimate weapon, such as the Death Star, has long been a sci-fi dream. And many of the weapons in fiction later became fact.

Leonardo da Vinci is well-known for his flying machine dreams, and his imagination also conjured up weapons such as a mechanical knight and a steam cannon. But it was the mechanical age of the nineteenth century when superweapons came to the fore, spawning an entire sub-genre of Future War fiction.

The atomic bomb, a precursor of the Death Star, was invented in sci-fi.

H.G. Wells had christened the very phrase "atomic bomb" in his prophetic 1914 novel, *The World Set Free*. His story led non-stop to Hiroshima.

Even though Wells was told by scientists, such as nuclear physicist Ernest Rutherford, that nature would "guard her secret" and atomic superweapons would never happen, Wells knew that atoms were the seats of enormous energies. They powered the very stars. And his book predicted a holocaust. The world's major cities are annihilated by small atomic bombs dropped by airplanes. This was no mere guesswork. Wells's weapons were truly nuclear; Einstein's equivalence of matter converted into fiery and explosive energy, triggered by a chain reaction.

And now the Death Star is the most famous science-fiction superweapon of them all. But what kind of energies would the Death Star need to annihilate Alderaan, or to vaporize the Earth?

That depends on something called the gravitational binding energy of the planet. This is the minimum energy you need added to a planet, or any other gravitationally-bound body, for it to cease being gravitationally bound. In other words, the energy needed to smash it into smithereens. It seems the gravitational binding energy the Death Star would need for either task, Earth or Alderaan, would be roughly the same.

Earth and Alderaan seem very similar planets. Alderaan is described an Earth-like world, four or five billion years old, covered with high snow-capped mountains and green pastures, which from space looks like a blue-green orb, covered in a web of white clouds. Sounds pretty Earth-like. Although we have no concrete evidence or measurements to support that the two planets are similar in mass, for the sake of the argument let us assume that the two worlds are similar in size based on their similarities, so that we are able to calculate the amount of energy needed by the Death Star to destroy Earth.

What's It Take to Destroy the Earth?

It's possible to conjure up an equation for this gravitational binding energy of a planet.

Start by considering something called the gravitational constant, G.

This is also known as Newton's constant, or colloquially as "Big G," and is one of the universal constants of nature. G is used in calculations of the gravitational force between bodies, and in the case of this equation the binding energy is proportional to 3 times G. The binding energy is also proportional to the mass of the planet squared, divided by 5 times the planet's radius (the distance from the very center of the planet to its surface).

When we plug values for G (6.674×10^{-11}), Earth mass (5.97237×10^{24} kilograms or 1.31668×10^{25} pounds), and Earth radius (6,371 kilometers or 3,958.8 miles) into the equation, we know what energy a Death Star Commander would need to destroy the Earth.

It's 2.24×10^{32} Joules of energy, or 224,000,000,000,000,000,000,000,000,000,000 Joules. That's a pretty impressive looking bundle of energy.

To give some kind of ballpark appreciation for how much energy this is, it's two hundred sextillion times the energy in an average lightning bolt, twenty quadrillion times the energy released by a severe thunderstorm, or roughly half a billion times the amount of energy that was released when that famous asteroid allegedly crashed into the Yucatan Peninsula, resulting in the formation of the Chicxulub Crater.

In a more cosmic and relevant comparison, the total energy output of the Sun, each second, is 3.8×10^{26} Joules. That means the Sun would produce the equivalent of Earth's gravitational binding energy in 6.8 days. So, a week's worth of sunshine could destroy the Earth. There's just the small matter of conjuring up that amount of energy to do the job.

What About an Antimatter Bomb?

Antimatter might do it.

Sci-fi is positively bristling with antimatter. It dreams up vast quantities of the stuff.

Chief Engineer Scotty uses frozen anti-hydrogen as the primary fuel for the propulsion of the starship *Enterprise.* Physicists in Dan Brown's *Angels and Demons* manage to create enough antimatter to blow up the Vatican. Science-fiction writers have imagined antimatter galaxies and even an entire antimatter universe.

Antimatter is made of stuff opposite in all ways to ordinary material. The idea was first mooted by physicist Paul Dirac in 1930. The existence of the positron, or anti-electron, was confirmed two years later.

The Death Star superlaser could deliver an antimatter bomb.

Assuming the Death Star weapon were actually a targeting laser, it could direct the siting of a huge antimatter device into the heart of a planet like the Earth. True, antimatter cannot easily exist in our universe. And the Death Star would probably need tons of antimatter, whereas only trillionths of a gram have been isolated in real labs.

But the potential energy of antimatter is colossal. It would make a fantastic power source, or a horrific bomb.

When antimatter joins explosively with ordinary matter, the result is 100% mutual annihilation. Since Einstein's famous equation $E=mc^2$ applies, a small amount of matter is converted into an enormous amount of energy. A mass of antimatter equivalent to a car could produce all of the world's electricity for one year.

So exactly how much antimatter would a Death Star commander need to get his hands on?

Around 1.24 trillion tons of the stuff. In terms of mass, this antimatter bomb would be about twenty thousand times less massive than the asteroid 16 Psyche, which we mentioned in our calculation of the cost of building the Death Star. 1.24 trillion tons of antimatter would make a sphere three kilometers across, about sixty times smaller than our asteroid.

That's a huge bomb.

But with a diameter of 120 kilometers, the Death Star could easily accommodate it. In fact, consider the relative size of the superlaser cannon well, the huge trademark dimple in the surface of the Death Star. The central hole of that cannon well is around six kilometers across, easily wide enough from which to project our antimatter bomb.

Think about the Death Star tractor beams. They could be used to project a force field that manipulated gravitational forces to push or pull an antimatter bomb. In Star Wars, such devices were often found on vessels, creating an energy field that allowed them to lock onto, and move, other vessels or objects.

"Wait, hang on," you might be asking yourself. "Wouldn't the bomb burn up as it passed through the Earth's atmosphere?" After all, when antimatter meets ordinary matter, the result is *kaboom*.

Here, the crafty Commander may consider antimatter option B. In this refinement, the Death Star could generate antimatter bullets of anti-Lucasonium—a super-dense material amassing a billion kilograms per cubic centimeter. This would be fired into the Earth's core. Lucasonium passes through ordinary matter as easily as a knife through butter. So the anti-Lucasonium bullet doesn't annihilate immediately. Instead, it accrues a protective sheath of plasma, as it plunges down to the Earth's core. Next up: a bullet of regular Lucasonium. This also falls to the core, at a time cleverly calculated to meet the first bullet head-on, at the very core of the Earth. In this instance, the two bullets annihilate each other and the Earth simultaneously. Not only is antimatter option B very space-efficient, but it also has the added advantage of releasing all the energy at the Earth's core, doing most damage.

Extinction Event

Death Star antimatter bombs wouldn't be the cause of the Earth's *first* mass extinction event. But it would certainly be its last.

There is still debate about the causes of mass extinctions in Earth's history. However, most scientists agree that large extinctions result when our biosphere undergoes a short-term shock. Of the major extinctions suffered by the Earth in its history, only one is associated with an extraterrestrial impact. That's the Cretaceous–Paleogene event that was a mass extinction of some three-quarters of the plant and animal species, including all non-avian dinosaurs.

Smaller impacts are not big enough to punch much of a hole in the extinction record.

Asteroids with a 1 kilometer (0.62 mile) diameter strike the Earth every 500,000 years or so. Smaller objects collide with us more frequently. Evidence of such impacts can be witnessed in the Barringer Crater in Arizona (a one-kilometer dent made by a nickel-iron meteorite only 50 meters [160 feet] across).

Larger collisions—with five-kilometer (three mile) objects—occur every twenty million years. The last known impact of an object of ten kilometers (six miles) or more in diameter was at the Cretaceous–Paleogene extinction event 66 million years ago. And we all know the havoc *that* caused. So the Death Star antimatter bomb would be the Earth's second "recent" extraterrestrial extinction event.

Of course, the Death Star Commander could consider other ways of destroying the Earth, other than the simple *zap-boom* antimatter option.

The Earth could be fissioned. This would need the Death Star to be some kind of universal fission machine, reducing every particle in the Earth down to hydrogen or helium. Alternatively, the Earth could be sucked into a microscopic black hole, though admittedly you'd need a microscopic black hole for that, and you'd need to be careful the Death Star itself wasn't also sucked in.

Luke Skywalker isn't the only potential pitfall a Death Star Commander faces.

WILL FUTURE MEGACITIES LOOK LIKE THE ECUMENOPOLIS OF CORUSCANT?

The central regions of the galaxy were among the most compactly populated and advanced worlds of the system. It could hardly help being the densest and richest clot of humanity the galaxy had ever seen. Its urbanization, progressing gradually, had at last reached a zenith. All of the planet's land surface was a single city. The population, at its height, ran into sheer billions. And this mass of humanity was devoted almost entirely to the administrative needs of Empire. Even then, they found themselves all too few for the complexity of the task. Daily, fleets of starships in tens of thousands brought the produce of pastoral idylls, known as the "summer planets," to the dinner tables of the City.

Its dependence on other worlds for food and, indeed, for all necessities of life, made the City Planet susceptible to conquest by siege. In the last millennium of the Empire, the monotonously frequent revolts made the City Fathers conscious of this, and Imperial policy became little more than the protection of the City's exposed and delicate jugular vein.

This vivid account at once seems like Galactic City, on Coruscant. Home to over a trillion souls. Built over thousands of years. Center of the human civilization of the Coruscanti.

And yet, the account is of an earlier ecumenopolis—a city made of the whole world. It's an account of Trantor, a fictional planet from the sci-fi Foundation series, written in 1942 by American author Isaac Asimov, he of the three robotic laws and *I, Robot*.

The word ecumenopolis (from the Greek οἰκουμένη, meaning "world," and πόλις, or "polis," meaning "city") had been coined in 1967 by Greek city planner, Constantinos Doxiadis. The idea was this: In time, the urban areas of the world would grow so large that they would eventually fuse. There'd be no city, as such. Just a continuous urban sprawl, worldwide. But, as usual, sci-fi had gotten there first—a generation before.

Cities of the Imagination

One of the architects of the skyscraper was Darwin.

No, not the beardy evolutionist. But his ingenious grandfather, Erasmus Darwin. One of his poems, *The Temple of Nature* (1802), foresaw a supreme science fictional vision: an overpopulated future of cars, nuclear submarines, and colossal skyscraper cities. Images of the future city soon became iconic.

Fritz Lang's *Metropolis* is a case in point. This 1927 sci-fi cinematic extravaganza, produced in Germany at the height of the Weimar Republic, was the most expensive silent movie of its day. Fritz Lang's stylistic and seminal work had been dubbed "Raygun Gothic." The film featured an architecture based on contemporary Modernism and Art Deco. But it was also a social commentary on design ideas for the modern-built environment—a futuristic dystopia of skyscrapers and social class.

Galactic City, Coruscant

Cue Coruscant, another city that spans an entire planet.

Here, too, is a futuristic dystopia of skyscrapers and social class. The surface level skyscrapers of Coruscant serve as homes and businesses for the rich, powerful, and politically savvy. The Coruscanti elites are ferried through the buzzing skylanes by private airspeeders, which weave from one tower to the next. Traffic is constant. However, crashes are rare, as speeders come equipped with auto-navigation systems that speed their travel pods along pre-programmed paths.

Among the Coruscanti elite are the Supreme Chancellor and members of the Galactic Senate. These rich and powerful people of the planet thrive in lavish lifestyles, lounge in (very) high-rise apartments, and dine in fine restaurants. Even as the Republic decayed, the Clone Wars raged on, and the average citizen became alienated, the Coruscanti elite continued to bask in their accustomed way, oblivious to the rest of the galaxy. Such are the rich. As befits an existence at the very highest levels, the elites breathe air that is clean and filtered.

Sunlight never reaches the lower levels.

The Coruscant underworld was a different story entirely. As Sheev Palpatine says to Anakin, "Do you know the rarest resource on Coruscant, my boy? Sky. Down here, the sun is a myth." It's a long way down from the highest surface on Coruscant, Level 5127, to the lowest, Level 1. Thousands of levels dwell beneath the city's surface structures, some deemed uninhabitable. The underworld has to be lit by artificial light, its inhabitants forced to breathe air toxic with the fumes of factory and vehicular waste.

The underworld houses Coruscant's criminal class. While much of the planet's blue-collar population lives in relative comfort, the underworld is a massive city beneath the city, accessible by huge portals. The many millions too poor to move upward, or hiding from the attention of the surveillance state, dwell in the bowels of the ecumenopolis. Such citizens rarely, if ever, see the surface of the planet during their lifetimes. They live at the mercy of violence, and the constant threat of death. It's a side of Star Wars that isn't as widely shown in its universe.

Future Megacities on Earth

How does the Coruscant ecumenopolis compare with future visions of terrestrial cities?

For the time being, it seems humans are committed to rather surface visions of the city. The future seems all chrome and chlorophyll. But perhaps these visions will quite quickly become as out-dated as Raygun Gothic. Prophecy is not a science. However, it's clear that the future will only contain what we put into it now. Rather than a sudden leap into dazzling Star Wars-style cityscapes, true creative innovation will unfold in real-time, as planners and architects respond to the sheer scale of what cities face in the future.

The new buzzword is survivability.

Our civilization relies on just the one planet. The Star Wars galaxy had other options. As with the account of Trantor in the beginning of this section, and like the terrestrial city-states of our ancient past, Star Wars planets could ship in produce from their own pastoral idylls, their own "summer planets." Earth is all we have at the moment.

Three quarters of our planet's major metropolises lie on the coast. Take China, the coming superpower of the twenty-first century. Each year, twenty million people migrate to Chinese cities, with the flood-prone Pearl River Delta now the world's largest urbanized area, according to the World Bank.

Worldwide, over a billion people in coastal cities will be vulnerable to serious flooding and extreme weather due to climate change by 2070, according to the *Guardian*. Millions more face domino effects, such as fresh water shortages, refugee vectors, and political instability.

Water Is the Watchword

The smart city of the future may be the floating city.

In the face of growing climate change, and rising seawaters, the first instinct of future city planners may be simply to take the defense approach. Build mega-engineering projects to keep the water at bay. In face of the flowing tide, retreat is not the only option. The city could be elevated above the water.

Water becomes the future datum. Rather than cities being a fixed and occupied space, they simply adapt to a fluctuating ground. The levels of Coruscant were many, and fixed. The future cities of Earth will be redesigned to accommodate rising water levels and adapt. Rather than keeping water out, the smart design will allow the water in. The border between city and nature, urban and rural, will be reconfigured, preventing cities from adopting the siege mentality of Trantor and Coruscant, and fighting a doomed battle with the elements.

In the making of *The Force Awakens*, Star Wars producers first considered the idea of Jakku as a water planet. Earth already is one. So, if you want a future vision of a terrestrial city, think less like Coruscant and more like Venice.

A city with water under its skirts. Flooded subways, submerged tubeways and sidewalks. Street level retail, underwater. Transport not by airspeeder, but by boat or dirigible. A slower and kinder lifestyle, a quieter type of city, without the constant thrum of the internal combustion engine. Rather than battling against nature, the futuristic city on Earth will welcome life on a planet consisting of predominantly water.

HOW LONG BEFORE WE BUILD INTELLIGENT MACHINES LIKE C-3PO?

It's one of the first characters we meet in the Star Wars universe and utters the first ever words of the franchise. C-3PO, the posh British butler bot—basically, a brass instrument with artificial intelligence and servo motors. Accompanying it is R2-D2, the cheeky but dependable swing bin on wheels, characterized by its beeps and whistles.

These robots entered our imaginations as droids with recognizable personalities. C-3PO, with its prissy and anxious demeanor, translated R2's utterings while providing a narrative to what was going on in the story. Constantly bickering with R2, they seemed like old friends, almost like an old married couple.

Although their structures were different, they both had a remarkable degree of artificial intelligence, being able to function autonomously and also exhibit loyalty. Robots like these could follow orders and break them if necessary, each following programming that suited its major function.

In the real world, though, when might we have intelligent robots like C-3PO?

The Rise of the Robot

The word "robot" was first used in Karel Capek's 1920s science fiction play, *Rossum's Universal Robots*. The word robot came from the Czech word *robota*, meaning "forced labor," referring to the featured characters that were thought of as human-like machines.

Since then, robots have frequently appeared in science fiction, from tiny nanobots to the huge Autobots. An iconic robot was Futura from Fritz Lang's 1927 movie *Metropolis*. Her design would find its way into the imagination of artist Ralph McQuarrie, who originally conceived of the look for C-3PO.

The robots of science fiction are often portrayed in a different way than the robots of real invention, though. Nowadays when we say robot we are normally referring to any programmable machine that can operate automatically.

Real robots have their biggest presence as industrial units for manufacturing. This is because they are extremely good at carrying out repetitive tasks, quickly and accurately. Some are static with maneuverable jointed arms, while others are mobile units that have freedom to move through a work environment.

Mobile robots have long since left the factory and are now being designed for use in hospitals, homes, toys, the military, and space travel. NASA has a space robot called "Robonaut," which is described as a "state-of-the-art, highly dexterous, humanoid robot." The advantage of being humanoid is in its ability to assist in human space activities.

Companies like Boston Dynamics have done much research into anthropomorphic and zoomorphic mobile robots alike, with names such as "Big Dog," "Cheetah," and "Pet man." Their latest anthropomorphic robot, "Atlas" has the motive capabilities of C-3PO, although without "3PO's" brass exterior and intelligence.

Kevin Warwick, a professor of cybernetics at the University of Reading, thinks that "we do have intelligent robots now but they tend to do one particular task and we're going to see that with, say, autonomous vehicles . . . on the road in the next ten or twenty years. It just depends how they're integrated. They will be intelligent in terms of that one particular task [but] with something like C-3PO it's multitasking. And he's a little bit human-like."

Built for Purpose

We're surrounded by machines. They're interwoven into the fabric of our everyday routines. They wash and dry our clothes, record our weekly television programs, and keep our homes at a comfortable temperature. To function effectively, each one of these devices requires suitable ways to interact with its potential environment.

For a mobile robot this could include sensors, wheels, jointed limbs, audio outputs, and some way to grasp things. It also needs a level of programming that can make "intelligent" use of it all. Beyond this, though, the level of a machine's intelligence only has to match the task it's required to do.

C-3PO was a protocol droid, mainly concerned with relations between sentient life-forms. In order to operate successfully in a humanoid environment it was given a humanoid form, which helps with opening doors, navigating stairs, and basically interacting in a world mainly designed for humanoids.

To aid in its role, it was fluent in over six million forms of communication and so would often function as a translator. It was also programmed for etiquette. Both of these functions require a thorough database of the cultural norms and customs of different species, as well as the ability to communicate in a way each species can understand. On Earth all vertebrates use sound as their main mode of communication, but alien species might use something more unique.

However, for a real world analog to C-3PO, Honda's butler bot ASIMO is probably the closest.

It's just over four feet tall and can operate for an hour using the thirteen-pound battery carried on its back. It can push carts and carry trays, and runs at almost four miles per hour. To gain awareness of its surroundings it has a three-meter range ultrasonic sensor and a two-meter range laser sensor for ground obstacles. It also has an infrared sensor for floor markings and two high-dynamic-range camera "eyes."

Its sensors and programming allow it to plot paths and detours while recognizing and avoiding objects in motion. It can also respond to audio received through its two microphones; turning to view novel sounds. Honda describes ASIMO as "the world's most advanced humanoid robot."

Humanoid bots like ASIMO maybe designed to look similar to us, but thinking like us is another challenge altogether.

Artificial Intelligence

Artificial intelligence (AI) is the theory and development of computer systems able to perform tasks normally requiring human intelligence, such as visual perception, speech recognition, decision-making, and translation between languages.

Maybe the most well-known test for AI is the Turing test. This is where a judge communicates with two unknown and unseen parties. One is a human; the other is a machine. The judge has to determine which is which by simply posing questions and judging the responses. If the judge can't tell the difference, then the machine has passed.

In 1950, when Turing proposed the test, computers were in their infancy. Examples such as UNIVAC required thousands of vacuum tubes to function. These room-sized machines required a bunch of people to operate the machines using punch cards and switches for data input.

Years later, with the miniaturization of components, their improved number crunching abilities have enabled more, complex interactions.

In 1996, world champion chess grandmaster Gary Kasparov was pitted against an IBM supercomputer called Deep Blue. He won four games to two. A year later he was defeated, three and a half games to two and a half.

In a more recent "man against machine" challenge another IBM machine called Watson was able to beat contestants on the game show Jeopardy. This was a breakthrough in that it could "understand" natural language and use evidence-based learning to deliver answers. This technology has now been included in a new educational smart toy called CogniToys.

Intelligence That Reads Between the Lines

C-3PO would need advanced levels of these functions to successfully provide relations between different species, such as Hutts and Humans. It would need to possess the ability to grasp the particular nuances of communication between life-forms, and then be capable of selecting between different possible meanings and choosing an appropriate response.

Crucially, AI can be integrated into any type of device where the ability to interact with humans is necessary. In the case of the AI called HAL from *2001: A Space Odyssey* it was a fundamental part of the running of a whole ship. We may see such AI entering the homes in the future, too.

Warwick points out that "we have pressures in terms of the aging society, so we may well need such machines pretty quickly to look after older people. So it doesn't look like C-3PO but it's a home with all sorts of sensors involved so it knows what the person needs."

In many ways these machines would have an advantage over humans in that they have a greater range of senses. The trick lies in the ability of the machine to recognize that a person may be having problems and respond accordingly. It would have to "understand" the individual's requirements and desires to function in a beneficial way. As such, this may require the AI to have some degree of empathy.

Some researchers have looked into empathic computing, describing it as "an emergent paradigm that enables a system to understand human states and feelings and to share this intimate information." They stress that it's not limited to helping an aging society. So there's a possibility that we may see empathy become part of a machine's capacity for "intelligence."

Intelligent Machines Like C-3PO

The future will see machines becoming much more dynamic in their programming, able to interact with humans on an even more personal level.

We've given them the means to observe us, and as they improve, we hand over more and more of our everyday mundane activities to them. We increasingly deal with machines for our shopping needs, as well as to automatically update us of our everyday affairs.

The programming that guides them has advanced machines from things that just respond to inputs, to entities that increasingly learn from situations and adapt. According to Warwick, "I think this learning and adaptability is an important part of our intelligence. Looking at intelligent machines . . . a lot of machines do learn nowadays . . . learning is a critical part of intelligence and it gives an individuality to each entity."

We're living in an age where smart technology surrounds us. We can talk to our device and it recognizes our words; software even allows rough translations between languages. You never know, we may just see the intelligence of C-3PO appear in a future upgrade to Siri.

When it appears, it may not look like a classic robot of science fiction, but it'll be just as remarkable.

WHAT ARE THE POSSIBLE FUTURE SOURCES FOR A TRACTOR BEAM?

The tractor beam. Once this invisible tether has you caught in its grips, there's no escape. Even the *Millennium Falcon* with all its capabilities could do nothing to overcome its awesome attraction.

The tractor beam was first coined by E. E. "Doc" Smith as a shorter version of the attractor beams from his classic Skylark series of stories. The attractors were described as beams of force that could be used to pin a prisoner to a wall, or to urge and lock two ships together.

Since then, many tractor beams have appeared in science fiction from alien abduction stories to methods for getting ships safely on board a mother ship. You might be wondering: How close are we to having actual tractor beams and how could they work?

The Tractor Beam

You remember the scene: The *Millennium Falcon* drops out of lightspeed and arrives at what they think is Alderaan. But it's not there, just rubble.

As they proceed, they're overtaken by a TIE fighter that appears to be heading to a moon. They quickly realize that it's no moon, but a space station. They try to change course to avoid the impending danger, but they can't. They're caught in a tractor beam.

It makes sense that a place like the Death Star would have a way of getting ships safely on board. Imagine leaving every landing to the skills of the pilots. A tractor beam seems like a great solution. It's kind of like having a tugboat to guide a large ship into a harbor. Except in the case of the *Millennium Falcon*, it was more like they had been caught by a hook and reeled in like a fish on an invisible line.

The tractor beam that snared the *Falcon* was not the only tractor beam on the Death Star. Not surprising, given that *Death Star I* had an estimated diameter of 120 kilometers, giving it a circumference of about 377 kilometers. It needed 768 tractor beam emitters located around its circumference to support its many docking ports.

The tractor beams had directionality, allowing them to lock onto nearby ships and objects. The beams were linked to a main control and coupled to the main reactors. This enabled Obi-Wan to sneak into the depths of the Death Star to sabotage the power source for the tractor beam holding the *Falcon*.

The Death Star wasn't the only vessel in Star Wars to have a tractor beam; however, it is the most notorious. Smaller ships like republic tugboats had two massive tractor beam projectors, while the Mon Calamari Star Cruiser had multiple projectors positioned along its 1.2-kilometer hull.

Invisible Attractors

So how could these invisible attractors work?

It's tempting to think that they might use some kind of tremendous magnet to grip onto the attractable parts of a ship. This has its problems, though.

Permanent magnets aren't extremely strong. They could be made bigger, but it's rather impractical. Electromagnets would be a more reasonable option but would still have to be unimaginably large or powerful.

Magnets interact with objects via their magnetic fields. As the distance increases from the source, the magnetic field gets progressively weaker. The field would have to reach out as far as the *Falcon*, which was pretty far out. Then it would have to be strong enough to overpower the sub-light propulsion of the *Falcon*.

Magnetic fields aren't focused beams either. To be a true tractor "beam" the force field has to be held within a parallel beam, i.e., one that doesn't really spread out over its distance. This is known as a collimated beam and would allow the tractor beam to be directed.

Instead, magnetic fields propagate out in every direction from the source. This means without adequate shielding it would penetrate into the Death Star, too, attracting everything else on the ship that was magnetic.

Despite these problems, there are still some magnetic devices that are publicized as tractor beam-like technologies. One such device is under investigation by NASA, who have teamed up with Arx Pax—the guys who brought us the Hendo Hoverboard.

Their aim is to use Arx Pax's Magnetic Field Architecture (MFA) technology "to create micro-satellite capture devices that can manipulate and couple satellites from a distance." It will basically be used to create a magnetic tether between their small four-inch sided CubeSats. However, these tethers will only work over distances in the order of centimeters.

Therefore, magnets are a bit of a long shot, and only work on magnetic metals. However, if we're looking for another invisible force that can attract any object then we need look no further than gravity. It's unrelenting and all-encompassing in what it draws in. If gravity can do it, what's to say an advanced technology can't be developed to achieve the same effect?

Gravitational Attractors

According to Star Wars canon, a tractor beam works by manipulating gravitational forces to grab objects.

The Death Star and other Star Wars spacecrafts have an onboard gravity generator, which stops the troops and cargo from flying about inside. It's possible the tractor beam is just an adapted version of this same technology. Where does modern science stand on this matter?

According to Einstein's theory of relativity, gravity is caused by a warping of space-time due to the presence of matter or energy. The more matter there is, the more space-time is warped, thus increasing the experienced force of gravity.

With gravity, the more mass there is, the stronger the gravitational field. Compared to magnetism, the attraction caused by gravity is extremely weak for the amount of matter needed to produce it.

However, a tractor beam working in this way could warp the space-time in front of an object causing the object to fall towards the warped region. The problem is that as far as we currently know, gravity cannot be focused; it operates in every direction from its source. So both ships would get drawn into this warped region of space depending on the relative mass of each ship.

Therefore gravity is possibly not a good solution unless a way can be found to control it and shield it from particular directions.

So what does science fact currently have to offer?

Real Tractor Beams

How about a technology that grips things with sound waves?

It's a form of acoustic levitation that can manipulate small particles, liquids, and living things to move them through air, water, and tissue. It isn't actually a beam but you've probably already seen the main problem. This technology wouldn't work in the vacuum of space. However, it could be used inside a ship.

Asier Marzo from Bristol University, who is part of the team that developed the technology, admits that "sound needs a medium to travel through, so as long as the manipulation takes place inside the ship, it should be doable. However, acoustic forces are very weak, so acoustic manipulation inside a spaceship would be more useful for orientating nuts or keeping droplets of liquids from floating adrift."

It only works on tiny objects, applying more to the microscopic realm, but it's still an effective tractor mechanism. It has a range of about seven centimeters, but depending on the power and aperture of the device this can vary. He hopes that in the future they may be able to manipulate objects up to a kilometer away using other types of pressure waves, such as air vortices.

What about tractor beam technology that does work in the vacuum of space?

Researchers at the University of St. Andrews have developed a light-based "tractor beam" to move microscopic particles. The light beam gives the same effect as when a solar sail is pushed along by light photons, however, it can operate in reverse by pulling or attracting the particles.

It works in liquids and in a vacuum like space. The problem is that the technique couldn't be used on a larger scale object, as it involves a transfer of energy. The amount of energy needed for a large object would create too much heating of the object. So it might be able to pull some dust off the *Millennium Falcon*, but that's about it.

In 2011, NASA scientist Dr. Paul Stysley set out to investigate and compare the viability of different laser-based tractor beam technologies, such as optical tweezers and Bessel beams. His team concluded that only solenoid beams, vortex pipelines, and optical conveyor belts were suitable for continued development efforts.

Before you get excited, though, these techniques have only demonstrated the possibility of moving tiny spheres forwards and backwards by a matter of tens of micrometers. But don't lose hope; it's still in its early days yet.

We might not have practical tractor beams like in Star Wars, but technologies have arisen that allow manipulation at a distance in ways that could only previously be seen in science fiction. Once again, scientists are rising to the difficult challenges of inching towards the technologies of our wildest imaginations.

COULD YOU REALLY DODGE OR DEFLECT A BLASTER BOLT?

So you've just landed on Cloud City and your old pal Lando invites you for a bit of refreshment. You follow him to a room, but as you head through the door, Darth Vader's sitting there waiting for you.

You react quickly, pulling your blaster from your hip and firing off a few shots in rapid succession. But Vader isn't fazed.

He blocks every single shot . . . with his hands.

Obviously Vader has some immense knowledge of the Force, and blocking blaster shots with his hand is pretty awe-inspiring, but being able to deflect blaster fire is a particular skill available to all Jedi and Sith.

So how possible is it for an average person (like ourselves) to avoid being shot by a blaster bolt?

Blaster Bolt Speed

Blaster bolts streak across the screen with a telltale glow, but to work out if a blaster bolt can be dodged, the first thing we need to know is how fast it's going.

Star Wars canon describes the blaster bolts as cohesive bursts of light-based energy. The blasters don't beam out light, though. They fire bolts of plasma energy, apparently by converting energy-rich gas into a glowing particle beam. As such, we can rule out lightspeed blaster bolts.

How about comparing a blaster bolt to tracer rounds?

Tracer rounds are bullets that commonly give off red or green blaster bolt-like streaks of light as they move through the air. They're mostly used by the military to help them see where they're shooting, but people often also use them on shooting ranges for their own personal entertainment.

When observed over sufficient distance, it's easy enough to follow the streaks of light. However, at shorter distances they move so fast that they appear more like streams of light. Most tracer bullets travel at a speed faster than sound, but some are slower. Either way, even the slowest tracer bullets seem faster than a blaster bolt. So we'll assume the blaster bolts travel slower than the speed of sound. Without an official document of blaster bolt speed, we'll have to resort to speculation. However, some scientists have gone to great lengths to analyze Star Wars footage to try and work out the speed that way.

Physics professor Rhett Allain came up with a figure for his online blog. It was roughly fifteen meters per second. On the other hand Adam Savage, from the popular science show *Mythbusters*, came up with an average figure of between 130 miles per hour and 135 miles per hour (about sixty meters per second). He even set up an experiment to see if a person could dodge a projectile moving that fast from twelve meters away. He didn't even come close to dodging it in any of the tests.

For the sake of our argument, we're going to assume a bolt speed of one hundred meters per second.

Reacting to a Blaster Bolt

So you're on Geonosis with a bunch of Jedi and you're surrounded by Count Dooku's army of droids.

Yoda appears in a ship leading a battalion of clones, and then all hell breaks loose.

Blaster bolts are flying everywhere, so you raise your lightsaber and instinctively motion to start blocking the shots.

The droids are twenty meters away from you. They fire their first shots.

As the bolt leaves the weapon, you're only made aware of the event once the light has reached your eyes. As a reference, light could travel around the Earth twice by the time you've blinked once. In this case the light takes fifteen millionths of a second to reach your eyes.

So the light has reached your eyes and you need to respond. But first you need to process the information through your central nervous system.

Our biology puts limits on how fast we can physically respond to a stimulus. For example, our visual response times can be about 0.19–0.25 seconds, whereas our response to audio stimuli can be quicker, coming at approximately 0.16 seconds. It takes longer to process visual sensory information because it involves a more complex cognitive system than our audio senses.

In the midst of a loud and busy battle, the visual stimulus would be the most important for a regular human. Once the blaster bolt had been perceived, a course of action would need to be decided by the brain. After that, signals would then have to be sent back to the body to initiate a response. Signals can travel through the nerves at a speed of up to 120 meters a second, so muscles can respond in just a hundredth of a second.

Due to the time it takes to process sensory information, the brain needs about half a second to consciously become aware that an event has happened. For example, in baseball, a player has about 0.4 seconds to respond and hit a ball after it leaves the pitcher's hand. The fact they achieve this at all is amazing, in that it's not actually enough time for their brain to comprehend that they've actually hit the ball. The batter only becomes consciously apparent of it about a tenth of a second later.

Of course a trained Jedi doesn't have to worry about such problems. The difference between a Jedi and a normal human is their Force sensitivity, in addition to the ability to move their limbs quicker. When Qui-Gon Jinn first discovers Anakin, he remarks: "He can see things before they happen. That's why he appears to have such quick reflexes. It's a Jedi trait."

Jedi Reflexes

It's something Jedi are born with and must train to improve. When Luke Skywalker first has a chance to train his lightsaber reflexes, he struggles.

The bolts are coming at him too fast for him to react.

Obi-Wan tells him that his eyes can deceive him, and that he must let go of his conscious self, relying on his instincts. On his next attempt, his eyes are covered but he does better, showing that a Jedi isn't using visual senses to react to the blaster bolts.

Using their sensitivity to the Force, Jedi can sense stimuli that normal humans are not even aware of. Without the Force, the ability of a Jedi to hit a blaster bolt with a lightsaber is as impossible as swinging a sword to successfully deflect a bullet. That being said, there is a human who has actually demonstrated a similar ability. He uses a samurai sword to split a tiny pellet shot from a BB gun. The pellet is four thousand times smaller than a baseball.

We return to swordsman Isao Machii, who was mentioned briefly in the chapter "Could We Ever Become Jedi?" He is a Japanese Iaido master who holds many records for his skills. He manages to cut the BB pellet, which is traveling at roughly 350 kilometers per hour (97 meters per second). He can also cut a tennis ball traveling at 820 kilometers per hour, which is 4.8 times faster than the fastest baseball pitch ever recorded (169 kilometers per hour).

Just like the Jedi, he had to train for many years to develop his expertise. However, he proves that a tiny object traveling at almost one hundred meters per second can be successfully intercepted by a human with a blade. However, for Machii to have thought about what he was doing would have taken about three-tenths of a second, by which time the pellet would have flown past. Dr. Ramani Durvasula of California State University describes Machii's skills as follows:

> *"This is about processing it in an entirely different sensory level because he's not visually processing this. This is a different level of anticipatory processing. Something so procedural . . . something so fluid for him."*

By establishing the highest response time of human beings, we've determined that dodging a blaster bolt would take more than simply reacting;

it's a form of preempting a strike, an anticipatory response or reflex that Machii seems to possess, just like the Jedi. The question arises: If Machii had a lightsaber, could it be used to actually deflect a bolt?

Deflecting the Blaster Bolt

The ability to deflect blaster bolts is an amazing feat performed frequently by the Jedi and Sith. They typically use their lightsabers to deflect the oncoming blast, except in particular case of Darth Vader, who has the ability to use his hands to the same affect. (Note: Kylo Ren—per his ability to control a blaster bolt in mid-air—and other powerful Force users, such as Yoda, could also possibly be grouped into this exceptional category.)

We've seen blaster bolts get repelled by magnetic seals and deflector shields, so anything that can replicate one of those should be able to also repel a blaster bolt. The key factor here is the ability of a magnetic field to deflect plasma.

If a blaster bolt were to hit a magnetic field, such as the one produced around a lightsaber, could a blaster bolt actually be deflected? This is what Space Plasma Physicist Martin Archer had to say on the matter:

> *"At large distances, it would simply be the external magnetic fields of the blaster bolt and the lightsaber interacting. This could deflect the bolt by a bit before it's even reached the lightsaber—depending on the orientation of the magnets they could, in contrast, attract the bolt. This probably wouldn't be a lot, though, given the speed of a blaster bolt and how magnetic fields drop off with distance, but under the right circumstances it could be enough to push the bolt to the side enough to miss you."*

So with a keen intuition it could be possible to deflect a blaster bolt with a lightsaber. If it's moving around one hundred meters per second, we could imagine a skilled swordsman like Machii achieving such a feat. However, for the rest of us, sadly without such superhuman abilities, we're destined to just get blasted by the bolt.

HOW LONG WILL IT BE BEFORE SMARTPHONES SEND HOLOGRAPHIC MESSAGES?

There's no doubt that the sight of R2-D2 beaming out a holographic message of Princess Leia is one of the most iconic visions in Star Wars.

They also have handheld holographic communicators, which is a pretty fantastic idea. (Imagine the drain on the battery, though.)

But could we ever have a smart phone with such a capability?

Holograms

In 1948, Dennis Gabor coined the word "hologram" from the Greek word *holos* meaning the whole, because it contained the whole information. Actual holograms, made using Gabor's 1947 invention of holography, would not be produced until the 1960s though, with the advent of the laser.

Jason Arthur Sapan, aka Dr. Laser, is the founder of Holographic Studios, New York. He's been working with holograms for four decades now. He says "they contain information about the whole picture . . . but each piece is slightly different from every other piece since it has its own view, which has a slightly different orientation."

Holographic images are stored within a film that records the interference pattern of incoming light, usually from a laser source. The image itself isn't in the film, but rather a representation in the form of the interference pattern. This interference pattern is the result of slight differences in the positions of the light waves when they enter the film.

When light is shone back onto the holographic film, it reflects back with the same configuration of light waves that originally created it, i.e., we see the original image, and can view it from as many angles as the original light hit it from.

The Star Wars holoprojectors don't appear to work that way, though. Instead they display an image within a 3D space, without the use of a holographic film. As such, they don't appear to be holograms at all. Sapan says, "Who is to say whether they are based on the principles of light wave interference or not? The real issue is that they are not what holograms are today, but science and technology are constantly growing and changing"

So Star Wars holograms may not be holograms at all. But then again, neither are most things passed off as holograms these days.

Ghostly Reflections and 3D Imagery

Many companies claim to create 3D images but really they are just 2D images reflected off a mirror or transparent solid surface. These "holograms" make use of an effect known as Pepper's ghost.

Pepper's ghost is a reflection-based holographic effect developed in the mid 1800s by scientists John Pepper and Henry Dircks. It was more recently used to make rapper Tupac Shakur appear on stage at Coachella Valley Music & Arts Festival in 2012—sixteen years after Tupac's death.

Other "holographic" techniques involve the use of a curved mirror to give the illusion that an image is suspended in mid-air in front of us. Sometimes referred to as a "reflecting hologram," this virtual image is due to the way the light rays from an original image reflect off of a concave mirror. This creates a three-dimensional real image, if the object being viewed is actually 3D. OPTI-GONE International has been using this technique since 1977 to create their popular Mirage toys.

In 1991, a concave mirror was also used in an arcade game called *Time Traveler,* advertised by Sega as a hologram. A similar method has also been used in Realview's interactive live holography, which they describe as "probably the most advanced 3D interactive visualization system." They say that "it accurately reconstructs points of light in mid-air."

However, even though the images it displays are 3D, they are actually non-holographic.

There is another class of display called volumetric displays where 3D graphics are displayed within a true physical 3D volume. Some are static volume displays like the 3D Arduino LED cubes. Others are swept volume displays, where different images are displayed at different positions on a surface that translates or rotates rapidly. Voxiebox uses this technique, as do 360° light field displays.

Despite the charm of these various displays, they still aren't what we think of when we imagine the Star Wars technology. But there is a new hope.

Plasma Projections

In Japan, researchers have come up with tangible holographic plasma, which as its name suggests is safe to the touch. They described it as "Fairy Lights in Femtoseconds."

A laser pulses in the order of tens of femtoseconds (one millionth of one billionth of a second) to create voxels (3D pixels) in mid-air. The voxels are shut off within seventeen milliseconds of a person touching them, which is well below the harmful exposure time of two thousand milliseconds (two seconds).

It's still not a hologram, though. Sapan says "it's a plasma display where the heat of the laser is vaporizing the air and creating a plasma point that lights up. The only use of holography here is in a HOE (holographic optical element—where a hologram becomes sort of a lens)."

Nonetheless, it's the closest thing anyone has ever come to producing Star Wars like "holographic" technology. The aerial and volumetric graphics rendered by the fairy lights technology can be scaled to produce sizes similar to some of the holograms that we see in Star Wars.

Yoichi Ochiai, from the University of Tsukuba, was one of the lead researchers on the technology. Regarding the range, he explained that "it depends on the size of the lens in the end part of the circuit. If we changed that lens we could project almost ten meters or twenty meters above."

So R2-D2 projecting the missing part of the map at the resistance base on D'Qar could be possible with this technology. However, at the moment, Dr. Ochiai claims that so far they have only projected to a size of twenty centimeters by ten centimeters; it just depends on the components of the optical circuit. He sees the future of the technology as being used for public viewing such as in a concert hall or even displaying information above an athlete in events like the Olympic Games.

It is also possible to use the technology as a laser scanner. It could capture 3D shapes such as humans, as well as recording some material properties of an object. So, the future is looking promising—but what about capturing more than just the front of an object, as is regularly done through the holoprojectors in Star Wars?

Image Capture

In Star Wars the images are fully three-dimensional, allowing the subject to be viewed from every angle. Considering in some cases the holographic unit is sat in front of the viewer, the problem becomes, how do we get the image of the subject from all angles?

At the moment, we could only speculate on such a technology.

When Anakin goes on a rampage in *Revenge of the Sith*, Yoda and Obi-Wan watch the events unfold on a holoprojector. We see Jedi entering and leaving the field of view. For this to work, the holoprojector would have to use some kind of sensor that can register a subject's position and features within the particular field of view.

There is an existing sensor that is remarkably close to this technology.

The Leap Motion controller can sense the position of any object put within a region equivalent to the shape of an inverted pyramid. The interaction space is about eight cubic feet (two feet above the sensor, and two feet deep on each side and two feet wide on each side). However, the company is slowly improving the range and scope.

The sensor uses cameras to track infrared light. As is, it can't see through objects, although the software can interpret the 3D data to infer positions of occluded objects. It's possible that a more advanced version of this exists within the Star Wars holoprojectors—one that can infer what things may look like on the hidden sides, and maybe even go further to incorporate information from other sources to complete the image.

So the question arises: Could any of this technology appear on smart--phone-like devices?

Smartphone Generation

Smartphones have come a long way since the term appeared in 1995 as a description of AT&Ts Phonewriter Communicator.

They now have high-definition color displays, high-definition cameras, and wireless networking on most standard devices. Some even have internal memory storage up to 128GB, and such novelties as glasses-free 3D displays and two cameras for filming in 3D.

There are even projector phones. The world's first debuted at the 2009 International Consumer Electronics Show in Las Vegas. More recently, Lenovo has advertised a laser-projector smartphone, called the "Smart-cast," which can project images to transform any surface into a fifty-inch touch display. This requires it to observe what it has displayed in order to allow it to act as a touch screen.

When it comes to true projected 3D images, though, we'll have to wait a while for further miniaturization of technology. Although optical components can be packed into really small sizes and compartments, there are still limitations inhibiting holographic technology being made available on smartphones. According to Dr. Ochiai, "Femtosecond laser sources are smaller now . . . almost fifty centimeters by sixty centimeters are now available. However, it's not enough power to generate a plasma."

Based on his research and expertise, Dr. Ochiai thinks that in twenty years they'll be a lot smaller, with our main obstacles being the power of the laser and the energy consumption. In regard to holographic projections from our smartphones, however, both he and Dr. Laser feel that that's something beyond the horizon of our current knowledge.

PART V
BIO-TECH

CAN YOU REALLY SURVIVE BEING FROZEN IN CARBONITE?

As Han Solo knows, there's more than one way to crack time travel.

First, there's the Einstein way. Professor Einstein showed that if you travel in a spaceship long enough, at speeds close to that of light, you can return to Earth, centuries in the future. Would-be time travelers are still waiting around, living life second by second, for such a spaceship to materialize.

Second, there's the cryonics way.

The method of choice for cryonics is not the spaceship, but the fridge. Clinical medicine is now able to switch people off for a period of time, leaving them with no heartbeat or brain activity. Used for certain surgical procedures, those who have been frozen say it was as if time stopped . . . and re-started an hour later.

Accident and heart-attack victims are brought back from the dead on a daily basis through the use of defibrillators and CPR. Patients' bodies are often cooled by neurosurgeons, so they can operate on aneurysms (enlarged blood vessels in the brain) without harm or rupture. Babies are born from the uterus of a mother whose embryos are frozen in fertility clinics, defrosted, and safely implanted.

Cryonics

Cryonics is ambitious. Its ultimate aim is the science of preserving human bodies in extremely cold temperatures with the hope of healthily reviving them sometime in the future. The notion is that, if someone has "died" from a disease that is incurable today, it may not be incurable in a brighter tomorrow. So the patient can be "frozen" and then revived in the future when a cure has been uncovered. A person preserved this way is said to be in cryonic suspension.

To grasp the concepts behind the cryonics, witness the news stories about people who've fallen into a frozen lake, and been submerged up to an hour in the icy water before being rescued. Those who survived did so because the freezing water put their body into a sort of suspended animation. The experience slowed down their metabolism and brain function to the point where they needed almost no oxygen.

Many futuristic fantasies using the cryonics theme were inspired by the ancient Egyptian habit of mummifying the dead. It was a relatively small imaginative step to suppose an arcane mummification process that preserved life and beauty, such as Egyptian princesses ripe for revival, may also feature in our future.

Famous kinds of cryonic suspension include that of Woody Allen in *Sleeper*, the 1973 American sci-fi comedy film, where he is frozen in time, and wakes defrosted two hundred years later in an ineptly run police state. It's also what happens to Fry in *Futurama*, the American animated sci-fi sitcom. Fry is a pizza delivery boy who, during the first few seconds of the year 2000, falls into a cryogenic capsule while delivering a pizza to Applied Cryogenics. He defrosts on the last day of the year 2999, and meets Leela, a one-eyed cryogenics counselor.

Of course, this is what happens to Han Solo, too. During the Galactic Civil War, Darth Vader blackmails Lando Calrissian into letting him use a carbon-freezing device on Han. Solo survives the ordeal, but it could have ended up a lot worse.

Carbonite

So, what exactly is carbonite?

According to the Star Wars canon, carbonite is described as a liquid substance, made from carbon gas, which could change into a solid through rapid freezing. Firstly, carbon is an excellent choice. We've written elsewhere in this book (see: "Would an Exogorth Really Evolve on an Asteroid?") about carbon's cosmic variability, its capacity to bond with life's other main elements to form the backbone of biology, and the fact that it forms the majority of all chemicals known to man.

The Star Wars account of carbonite suggests that the substance was used before the invention of the hyperdrive, with early spacers (a slang term for someone who spent a large part of their life in space) using carbonite to endure long journeys. But the technique had brutal side effects, known as hibernation sickness.

Little wonder that bounty hunter Boba Fett was worried Han Solo would be *iced* for good when encased in carbonite. The freezing process might "ice" his prize in the mafia sense of the word, as well as the literal one.

The idea of using a deep freeze to travel through time was explored in Jack London's first published work, *A Thousand Deaths* (1889). In this, the hero is repeatedly killed and brought back to life by his mad scientist father:

> *"After being suffocated, he kept me in cold storage for three months, not permitting me to freeze or decay. This was without my knowledge, and I was in a great fright on discovering the lapse of time."*

From that point on, extreme cold has been used throughout science fiction to help heroes, or villains, travel through time.

Don't Try This At Home

Turns out Boba Fett had good reason for concern.

On Earth, of course, it's illegal to perform cryonic suspension on someone who is still alive. So please don't try it at home. Terrestrial research shows that freezing damages the structure and integrity of tissues, so that stuff leaks out when the temperature is raised again. This much you *can* try at home: freeze and defrost a strawberry and you will soon have a good idea of the mushy mess that Han might become.

So, even though a person may appear preserved on the surface, on a cellular level the damage is disastrous. Nature boasts a few creatures that can weather the freezing process—some frogs, fish, turtles, and insects. Their bodies produce huge amounts of the carbon-based sugar glucose, a natural antifreeze preventing ice-crystal formation. Sadly, sufficient levels of glucose in humans would come with a side-effect of death.

But the terrestrial version of "freezing" folk reveals some possible secrets about what might have happened to Han.

Once transported to a cryonics facility, patients aren't simply slammed into a vat of liquid nitrogen. The water inside their cells would freeze. When water freezes, it expands. Their cells would simply shatter. The cryonics experts must somehow remove water from the cells and swap it with cryoprotectant, a sort of human antifreeze.

This cryoprotectant, a glycerol-based chemical mixture, is an attempt to make for humans a version of the natural antifreeze that frogs and fish eke out with ease. The aim is to shield organs and tissues from the formation of ice crystals at very low temperatures. The process of deep cooling without freezing is known as vitrification, and puts the cells into the kind of suspended animation formerly dreamt of in fiction.

Freezing People Has Its Problems

Now comes the tricky bit.

Once its water is replaced with cryoprotectant, the body is cooled on a bed of dry ice until it reaches -130°C (-202°F), and vitrification is complete. Next, the body is placed in a container, which is immersed in a large metal tank filled with liquid nitrogen at a cool -196°C (-320°F). The body is stored head down, and for good reason. If there were ever a leak in the tank, the brain would stay immersed in the freezing liquid.

Unlike Han, Earthlings in cryonic suspension haven't yet been revived successfully.

But cryobiologists hope that a new technique, called nanotechnology, will soon make revival a reality. Nanotechnology uses microscopic machines to manipulate single atoms—the building blocks of all things, even organisms—to build or repair human cells and tissues. Futurists hope that soon nanotechnology will repair not only the cell damage caused by freezing, but also the damage done by aging and disease. Some cryobiologists predict that the first cryonic revival might occur somewhere around the year 2040.

Your Local 24/7 Cryonics Store

If you should meet your demise while reading this book (granted, that's cosmically and statistically unlikely), you could always try cryonics out for yourself, of course. Try, by choice, something that Han Solo had no choice in trying.

Cryonics is big business, but it isn't cheap.

It can cost up to $150,000 to have a whole-body preservation. For more frugal futurists, you can preserve just your brain for a cool fifty thousand dollars—an option known as neurosuspension. Keep your fingers crossed that, should you opt for this brain treatment, vitrification technology will come up with a way to clone or regenerate the rest of you.

The cryonics industry is currently working on chemicals to perform the carbon-freezing function with less death. Han Solo was presumably pumped full of these. The good news is, once the freezing process is done, a study undertaken by cryobiologists (recently published in the *Rejuvenation Research* journal) showed that memories are preserved, at least in worms (see: "Would an Exogorth Really Evolve on an Asteroid?").

THE AGE OF C-3PO: HOW WILL EARTH DEAL WITH THE DROID ARMY?

The Age of C-3PO is upon us.

Droids existed in Star Wars since the early days of the Jedi Order. In the coming decades of the twenty-first century, the real world will likely see the revolutionary rise of robots. This revolution will transform private and public life just as radically as the Internet and social media have rebooted life in the last decade.

This is no mere cant. This expert opinion is that of Californian think-tank The Institute for the Future. The IFTF is the Earth's most respected think-tank. It's been plotting the course of the future for government and corporate concerns since it was cleaved off the RAND Corporation in 1968. The IFTF says droids will increasingly dominate the twenty-first century—from the way we wage wars to the way we spend our leisure time.

Stephen Hawking, previously Lucasian professor of mathematics at Cambridge University, says we face an "intelligence explosion," as machines will engineer themselves to be far more intelligent than humans. Multinational entrepreneur, Elon Musk, has called the prospect of artificial intelligence "our greatest existential threat," and Bill Gates, who knows a thing or two about computer systems, admitted he is "in the camp that is concerned about super intelligence." Maybe all three intellects are fully aware of the Clone Wars.

But if robots and droids are the future of terrestrial work and war, where will humans fit in?

The Droid Army in Star Wars

Droids in Star Wars can be considered to fall roughly into two types, exhibits A and B.

Exhibit A: Star Wars leisure droids

Over time, droids became ubiquitous in the everyday life and operations of the galaxy. They were trusted with a legion of tasks, ranging from the running of basic diagnostic systems to performing complex surgical procedures. Droids were even trusted to fly starships. Depending on their function, droids were categorized into classes. Medical droids, for instance, were categorized as class one, security droids as class four, and construction droids as class five.

Exhibit B: Star Wars battle droids

More ominously, during the pan-galactic war (also known as the Clone Wars) the Confederacy of Independent Systems (CIS) employed a Separatist Droid Army, led by the infamous Kaleesh cyborg, General Grievous. The Army was comprised almost exclusively of droids and numbered in the quintillions. The troops were created and tooled up by Baktoid Combat Automata, a manufacturer of battle droids owned by the Techno Union, an unholy alliance of corporate giants who feigned neutrality in the wars, but in deeds actually armed and supported the CIS. When the Clone Wars ended, the Techno Union was absorbed into the Galactic Empire. Under the new regime, many people in the galaxy were afraid or otherwise distrustful of droids, due to their memories of bad droid behavior during the war. But what does planet Earth currently have to offer in the way of droids?

The Droids You May Not Be Looking For

We could similarly split terrestrial droids into two types, exhibits C and D.

Exhibit C: Terrestrial leisure droids

Terrestrial droids are already on the march.

They're making inroads into construction. A machine system, created by the University of Southern California, is able to construct buildings in layers, guided by droids. The system is a labor-shaving enterprise, too, reducing construction times and costs by up to 75%. Fancy a droid as a teacher? In the rote-learning region of South Korea, robots are used as teaching assistants. In language classes, students are made to parrot words and phrases uttered by the droid, and are assessed on how well they copy. Nowhere near as bad as cloning the convolutions of Yoda perhaps, but not exactly progressive, soulful teaching.

Next are the motor-droids. Motor-droids may not be astromechs, but they represent the most revolutionary change to terrestrial transport since the invention of the internal combustion engine. Google and other companies are working on driverless cars. Governments around the globe are investing in a future full of automatic road freight. These examples of motor-droids, which is essentially what they are, are forecast to make up 75% of all traffic by 2040.

The rise of the motor-droid promises not just of hordes of jobless drivers, but also of the transformation of the associated traffic infrastructure, from training to fuel stations. There's also the potential for motor-droid piracy. Star Wars is rife with pirates and bounty hunters. A future Earth of droid-driven traffic is full of the promise of pirates boarding freight-droids, or hacker-jackers, with the skills needed to hack into droid software, and hijack the freight onto an alternate journey of their choosing.

Inevitably the rise of the robots will put people out of work. Bank of America Merrill Lynch foresees 47% of US jobs will be *droided* out by 2020. The Boston Consulting Group predicts droid software or droids will replace 25% of jobs by 2025.This robotic revolution doesn't just affect America; the Australian Committee for Economic Development forecasts that five million Aussie jobs are set to disappear before the same year.

The key difference between our droid future and similar changes from terrestrial history is simply the utter pace of change. It's not just blue-collar workers who will be replaced by droids. A study by researchers at Oxford University and Deloitte predicts that 95% of accountants are likely to be automated by 2020. In Australia, that's 183,900 accountants that will go the way of the dodo.

The workless future will be similar to the mechanization of agriculture. People will have to learn to do other things, making way for the droids. It's becoming increasingly easier to imagine a near future where we begin to classify droids in the Star Wars way.

Machines for Military Means

Exhibit D: Terrestrial battle droids

The development of terrestrial battle droids is a double-edged endeavor.

One road is what we might call the Grievous Route—making living species into battle cyborgs. This approach involves tooling-up, or augmenting, the human body with robotic weapons systems. Manufacturers say this tactic offers "the best of both worlds," as it combines the rapid reaction times and precision of robotic systems with the superior cognitive abilities of humans.

The ultimate aim of some in the Pentagon is the development of "super soldiers." These cyborgs could be deployed anywhere in the world within hours, and stay in the battle zone for protracted periods. Cyborg body and function could be optimized through nano-sensors that continually monitor medical status, embedded nano-needles that decant drugs when needed, and even nano-robots that can swiftly heal wounds during battle.

The other road is the Droid Route—pure robot soldier.

The leading prototype here is the aforementioned "rescue" robot named "Atlas," a 6'2", 330-pound droid that moves like a human, but is designed to go to places humans cannot, such as nuclear reactors and raging wildfires. Or—when out of "rescue" mode—into battle.

Critics are concerned that droids could advance in undesirable directions, and in ways not predicted by droid scientists. With this fear foremost, military minds have—for now at least—suggested humans will never be completely removed from the droid decision loop. While droid weapon systems will become ever more autonomous, they may never entirely replace human cyborgs on the battlefield. At least, that's the plan.

The Age When Robots Rule

Exhibit E: the terrestrial Em-Droid

This third kind of droid, one not dreamt up in the stories of Star Wars, is the first truly smart terrestrial droid, one that will be based on brain emulations, or ems. Professor Robin Hanson, economist and scholar at Oxford's Future of Humanity Institute, thinks the droid revolution, when it comes, will be in the form of the Em-Droid. You take the best and brightest two hundred scientists on the planet (ironically this may include Hawking, Gates, and Musk), then you scan their brains and upload their consciousness into a robot. This is the Em-Droid—a robot indistinguishable from the humans upon whom they are based, except a thousand times faster and fitter for the future.

Train an Em-Droid to do a job and clone it a million times: an army of droids is at your disposal. When Em-Droids can be made cheaply, they will displace humans in almost every job. In the new Em-Droid economy, the global economic output may double every few weeks, and competition will drive nearly all wages down to subsistence levels.

Trillions of Em-Droids will live in tall, liquid-cooled skyscrapers in searingly hot cities. Given the nature of the Em-Droid, they will be very able and focused workaholics, who respect and trust each other more than humans do. Some Em-Droids will have bodies; others will simply live in virtual reality, as Em-Droid white-collar workers won't need bodies.

The Em-Droids will collect in related "clans" and use "decision markets" to make crucial corporate and political decisions. Em-Droids will work almost all the time, but choose to dream about an existence that is nearly all leisure. Surveillance will be total.

There is potential for a huge backlash.

How will humans react when told their consciousness will be uploaded into a droid, and their body destroyed? Isn't that murder? The Em-Droid clone of you won't actually be you. With such intelligence at their disposal, Em-Droids are also likely to crack the problem of proper artificial intelligence very quickly.

The rise of the Em-Droids is exactly the kind of thing that Hawking warns about in his phrase, "the intelligence explosion," and what Elon Musk means by, "our greatest existential threat."

It's time to join the Rebel Alliance.

WHEN WILL A CLONE ARMY BE BUILT ON EARTH?

EXT. TIPOCA CITY, KAMINO—DAY

The reality of the star system is a perfect clone of its star map hologram. Obi-Wan's Starship flies over the storm-shrouded planet of Kamino.

In heavy rains and hard-driving winds, he sets down on a landing platform in Tipoca, the vast, ultra-modern city on stilts, which sits in the ubiquitous waves and swirling rains of this watery world.

INT. TIPOCA CITY, CORRIDOR ENTRANCE

Obi-Wan walks through a sliding door into the brilliant white light of the high-tech cloning facility. Kamino is the birthplace of the Grand Army of the Republic—the legions of clone troopers, identical soldiers bred to serve and fight for the Republic.

Developed and grown by the slender, long-necked alien race of Kaminoans, the clones were based on the DNA of bounty hunter Jango Fett, another resident of Tipoca City.

EXT. GRAND ARMY FORMATION, TIPOCA CITY, PARADE GROUND (RAINSTORM)—DAY

Obi-Wan, along with Kaminoans Lama Su and Taun We, come out onto a balcony. Far down below is a huge parade ground. The rain and wind are brutal. Thousands of troopers, faces covered by helmets, are marching and

drilling in formations of several hundred.

We enter the mind of a single clone trooper, a "bad batcher" according to some.

BAD BATCHER

They used to say that a child conceived in love has a better chance of happiness. They don't say that anymore. Not to me. I was conceived here, in the driving rain of Kamino. Churned out, among a fifth of a million other clone soldiers.

Those that don't do as they're told, or think straight, are thought below par. "Bad batcher," that's me. Sick of seeing my own face, day in, day out, to the first syllable of extinction. I feel I belong to a new underclass of human. No longer determined by social class, or the color of skin. Nah, this new discrimination is all down to science.

Though we shed five hundred million cells a day, I'm the same person I was yesterday. Same as all the others. Almost. Hatched by the thousand, accelerated growth in half the usual time. Combat, training, combat, training. It's tedious. A life led just for the battlefield.

The soldiers of old would have life experiences to bring into battle. Subtlety. Nuance. Empathy. Things that would make them more able to understand how enemy soldiers might think, and act. But I guess there's so many different species in the galaxy, clone troopers would have a hard job working through all those many alien cultures anyway. Might as well shoot first, think later. Most mission tasks pit us up against droids in the Separatist army. But none of these jobs are for me. I'm no front line cannon fodder. Don't care if it's the orders of the Republic or the Separatists.

There's no gene for fate.

When the going is good, the good get going. So, I'm gonna make my escape, into another star system, another galaxy. For someone who was never meant for this world, I must confess I'm not going to have a hard time leaving it. They say every atom in a clone trooper body was once part of a star. Maybe I'm not deserting . . . maybe I'm going home.

Terrestrial Plans for Super Soldiers

Is life once more set to imitate art? Will Earthly politicians and their armies go the clone way of Star Wars?

The signs are promising. We're all aware of the droning news that there's no spare cash to go around, in this the most austere of ages. And yet, there's always money for war (maybe someone found the odd few billion or two down the back of the couch of an arms dealer).

Let's examine the Pentagon's DARPA (Defense Advanced Research Projects Agency) creation of a super soldier army. Super seems to be key here. We're talking a *super* army of *super* soldiers, and with *super* human abilities. It seems those military generals have been paying quite close attention to all those superhero movies. All this *super* stuff is to be achieved through the wonders of genetic modification. It seems the program has persisted for many years, shrouded in *super* secrecy (you have to instill some level of secrecy—that's another rule of the superhero movies the generals have been delighted by).

The generals are delighted, too, that these mutant soldiers will revolutionize future war.

Science will make the killing more efficient. The killing was pretty efficient in WWII. The Soviet Union lost twenty-seven million humans, China lost over fifteen million humans, and Germany and Poland lost around six million humans apiece. The total casualties came to over seventy million. But DARPA is striving for more efficiency. They believe the genetic modification of targeted human genes will give their super soldiers an advantage on the battlefield. This science will give rise to "the most amazing abilities and performances," and consequently, more efficient death.

Follow the Money

The naïve might argue that a main historic lesson in the art of war is that the enemy will soon match any advantage.

However, the generals are having none of that. Instead, their mantra bleats of the scientific advance of gene wangling. The super soldiers will be smarter, sharper. They'll be more focused, and physically fitter, than enemy soldiers. Some of the claims even sound like they're out of the Star Wars canon: the super soldiers "will be capable of telepathy, run faster than Olympic champions, lift record-breaking weights through the development of exoskeletons, re-grow limbs lost in combat, possess a super-strong immune system, go for days and days without food or sleep." That kind of thing.

Those skeptical of such Star Wars-sounding claims need only to follow the money.

DARPA has invested millions in preclinical research studies to develop a means of soldiers surviving significant blood loss. This "breakthrough" would solve the problem of needing life-saving medical treatment immediately after combat injury. That's clearly a strategic problem during the snags and hazards faced in the heat of battle.

Yet, more millions have been invested in squirrel power.

Yes, there's a squirrel gene that makes an enzyme in the pancreas, which enables hibernation through the winter months. Genetic modification (GM) research is being conducted into the possibility of taking the gene from the squirrel and implanting it into soldiers, allowing the world's first ever GM GIs! (Cue Youtube clip of "bad batcher" GIs who've taken on too many squirrel traits and are spied busily burying their nuts where no-one else can find them.)

Mind Control

Perhaps the boundary between science fact and fiction blurs most on the question of mind control.

There are plans to blunt the soldiers' brains. A forty-million-dollar grant has been handed out to develop memory-controlling implants. This may allow any remaining keenly edged emotional acumen to be tampered with. The aim is to delete the soldiers' empathy genes so that not only do they show no mercy, but are also devoid of fear. A disturbing, dark side-sounding "Human Assisted Neural Devices (HAND) program" will control soldiers' brains by a remote "joystick," which operates from some far away control center.

What will be the future cost of this blurring between fact and Star Wars fiction?

Are civilians next? You can imagine the Empire would be delighted to get its fingers into some of this technology. Memory-controlling implants sound like the perfect start to a scientific dictatorship, one that could set down false memories, as well as behavioral control programs.

But let's take a step back.

DARPA has a history of oddball technology. Sure, they've had astounding successes with far-sighted technological leaps, such as the first virtual reality devices, and a precursor of the modern Internet. But squirrel power? Little wonder that DARPA's seemingly madcap schemes have resulted in its reputation as the US military's "mad scientist wing."

We can't put a figure on when exactly the clone army will be built, but plans are being made.

WHAT EXACTLY IS THE FORCE?

It's that exciting element of Star Wars that makes you feel as if anything could be possible if you learn to master it. It's what gives a Jedi his power; it's the reason things can be levitated; it's responsible for mind control, and it even allows individuals to perceive the future.

It is, of course, the Force. But do any of these features even remotely correspond with scientific knowledge, or do we have to invoke Arthur C. Clarke's third law again? Namely, that "any sufficiently advanced technology is indistinguishable from magic."

What exactly is the Force?

The Unseen Presence

The Force enables all sorts of amazing and seemingly impossible things to be performed. It's not a straightforward entity to conquer or understand, though. Its intangible nature makes it something of an enigma, and also something that may have to be experienced to be believed.

If you want to get to grips with the Force, you must be one with it. Don't think . . . stretch out with your feelings. Well, at least that's what the Jedi masters say.

It's something that relies on your intuition as opposed to your eyes, which can deceive you. Through it you can see other places: the future, the past, and old friends long gone. In this scenario, the Force is something that acts on the subconscious processes of your brain, guiding you as a kind of extra sense.

Throughout history, there have been many claims of extrasensory perception (ESP). However, the first real experiments and popularization of ESP occurred in the 1930s with J. B. Rhine's research at Duke University. He was first to employ the card-guessing technique, which utilized Zener cards. Each card is printed with the image of a different shape, such as a circle, a square, a few waves, a cross, or a star. With the symbol side hidden, the person being tested for extrasensory perception is asked to guess which symbol is on the card, while the proctor records the results.

In *The Phantom Menace* we see the young Anakin Skywalker being put through a similar test by Mace Windu. He guesses all the objects correctly, demonstrating to the Jedi Council that the Force is truly strong with him.

In the real experiments of the 1930s, and in subsequent tests, there have been no conclusively successful candidates for ESP.

A Mystical Force

The Force incorporates ideas from many cultures such as the Chinese principle of Qi and the Roman Catholic God. The famous Jedi well-wish "May the Force be with you" is an obvious refiguring of the saying "The Lord be with you." So it's possible the Force is meant to be something more spiritual than scientific.

Perhaps the Force represents an all-powerful and ever-present God, with the miracles of Jesus replaced by the abilities of the Jedi. Despite the possible symbolism, this abstract theory would bring us no closer to knowing what the Force *really* is.

The problem is, at a quick glance, these miracles and tricks appear to have no real mechanism through which they can be reliably understood. How exactly could a Jedi be guided by some mystical force?

There is an ability in some animals that allows them to use an invisible force as a guide. It's called magnetoreception and it relies on the Earth's magnetic field. Animals like salamanders, frogs, and migratory birds use this mechanism to gain an added sense of direction.

For years scientists wondered how exactly this was done, considering that the geomagnetic field is so weak. Now, they have identified the mechanism in birds and, impressively, it involves their eyes. What's even more interesting is that Scientist Joe Kirschvink at California Institute of

Technology has been looking for a magnetic sixth sense in humans, too.

His tests showed that a set of neurons in the brain were firing in response to a varied magnetic field. The test subjects were not aware of the magnetic field, but their brains were definitely responding. The signal was picked up on an electroencephalograph (EEG). If a person could somehow tap into this response exhibited by their brain, they could be able to use the effect as a kind of sixth sense. In regard to magnetoreception, Kirschvink said, "It's part of our evolutionary history. Magnetoreception may be the primal sense."

Fields of Understanding

A huge change in human understanding came with the concept of fields of influence. Instead of just saying there is a force of gravity, we can say there is a gravitational field. When something with mass is put within that field, it can experience a gravitational force of attraction due to its interaction with the field.

In a similar way, electric charges have an electric field, which exerts a force on any charged object placed within the field. Likewise, magnetic materials have a magnetic field, which will exert a force on ferromagnetic materials (basically, materials that can attract to magnets). A magnetic field can also be found around an electric current (a flow of electric charge, such as electrons or ions). This is where the Earth's magnetic field comes from.

It's now known that the phenomena of electricity, magnetism, and light are all different aspects of the same field, called the electromagnetic field, which carries the electromagnetic force. This is just one of four fundamental forces of nature also known as fundamental interactions. The other three are gravitational, strong nuclear, and weak nuclear. Each one has an associated particle that is said to convey its force. These are known as force carriers.

In Star Wars, the Force is described as an energy field, although it's not a field that modern physics is currently aware of. Even though the Force does involve some of the fundamental interactions, scientists don't currently know whether there is one overarching force or framework that could make them all related. However, scientists have been searching for something like that; it's often referred to as a theory of everything.

Force and the Universe

The universe is everything that exists and science is the venture that seeks to understand its complex workings.

Roughly 68% of the universe is dark energy and around 27% is dark matter, both of which scientists are literally in the dark about. The other 5% is more familiar and includes all the atoms that comprise the things we see around us.

Atoms are extremely spacious. The central part, called the nucleus, is only about one hundred thousandth of the size of an atom. As such, more than 99.99% of your body and everything around you is actually empty space. The exception would be the fundamental forces, whose fields are able to mediate that "empty" space.

In Star Wars, the Force is said to "bind the galaxy together." This is true of the gravitational force, whose force carrier is thought to be gravitons. Its field arises from mass and also acts upon mass in a way that attracts it together. It holds planets and stars together, as well as holding stars together in galaxies, and galaxies together in clusters and super clusters.

At the atomic level, it's the weakest of all the fundamental forces. However, its strength is accumulative, meaning the more mass you have, the greater the force of attraction. Its field is also infinite, getting weaker with distance, but not disappearing. In 2015, scientists managed to detect gravitational waves emanating from a pair of merging black holes, 1.4 billion light-years away.

The Star Wars Force is also said to bind us together. It surrounds and penetrates the characters in the story, flowing between everything that they see. This is also true of a force in reality, but this time it involves the electromagnetic (EM) interaction.

Electromagnetism is the force that holds our atoms together, while also flowing through us as electromagnetic radiation. It's also the force that prevents your backside from falling through the seat, even though the atoms are 99.99% space. Its force is conveyed via photons—the unit of light.

We can see objects because photons travel to our eyes from said object. Photons can exist at lower energies, such as radio waves, all the way up to high-energy gamma rays. At distances and masses that are within our everyday experience, EM is much stronger than gravitation. The EM field is also infinite, as evidenced by the fact that we can see the light from galaxies that are billions of light-years away.

The other two fundamental forces are nuclear, meaning they essentially only exist within the nuclei of atoms. They both have very short-range fields, about a trillion times smaller than a millimeter.

The strong force is what keeps the nucleus together despite the protons having the same charge and wanting to repel. It's much stronger than EM force and its force carrier is gluons, which mediate force between the quarks.

Quarks are elementary particles, meaning they are not formed from smaller particles. There are six "flavors" of quark, two of which are used to form the nucleons (protons and neutrons) in the nuclei of atoms. Three quarks combine to form each nucleon. The proton contains two "Up" and one "Down" quark, while the neutron contains one "Up" and two "Down" quarks.

The weak force is about one million times weaker than the strong force. It's responsible for radioactive decay (and interactions with neutrinos). Its force carriers are either positively or negatively charged W bosons or neutrally charged Z bosons. Bosons are the broader family of which all force carriers are members.

In 2013, the Higgs Boson was confirmed. It's thought to provide some evidence for a Higgs field that gives mass to particles, while permeating the entire universe. Maybe this is a step towards identifying the true nature of the Force.

Despite us not being aware of a single force that could mirror the Force described in Star Wars, maybe it's just beyond our ability to measure it at the moment. Perhaps the Force exists, but as an entity so intangible that it lies beyond the scope of scientific knowledge; somewhere in the realm of Arthur C. Clarke's sufficiently advanced technology. We may just have to wait and see.

DOES THE LIGHTSABER REALLY CUT IT?

It's the weapon of a Jedi knight—one of the most recognizable weapons in science fiction.

Every kid who's seen Star Wars has pretended to wield one if not wished they actually had their own, despite their being way more lethal than any real sword could ever be. A definite safety hazard, but highly regarded nonetheless.

Jedi don't come about them easily either. As a youth, they must locate and retrieve the kyber crystal that will become the "heart" of their first lightsaber. Then, they have to build it themselves under the guidance of a tutor. This is part of the trials of a youngling towards becoming a padawan.

Seeing as these adolescent but hugely talented younglings can build a lightsaber, we have to wonder: How likely is it that scientists and inventors can make one in real life?

Props to the Replicas

The lightsabers of the original trilogy were made from the handle of a Graflex flash unit. Special effects specialist John Stears modified the handles to create the famous hilts that would be the source of the Jedi's glowing blade.

A few companies specialize in high-quality lightsaber replicas. They are commonly sold for hundreds of dollars, and are so good that they've even influenced the props used on set for recent movies such as *The Force Awakens*. One replica was sold on Ebay for more than fifteen thousand dollars. Imagine how much a real one would cost!

Lightsabers come in many colors, including green, blue, purple, white, or red (for the Sith). Each lightsaber is customizable. There are double-bladed lightsabers such as Darth's Maul's in *The Phantom Menace*, and versions with crossguards like Kylo Ren's in *The Force Awakens*. The owner is linked to their lightsaber through the kyber crystal, which resonates with the Force.

Some serious thought has been put into making real-life lightsaber-like devices.

Allen Pan of the Youtube channel Sufficiently Advanced has adopted an interesting idea. He modified a lightsaber prop handle to house flammable gas that can be propelled through a thin needle. It basically functions as a flamethrower but the thin blade-like flame glows like a lightsaber. The flame color can even be changed.

As compelling as the replica may look, it still can't really cut through things, and being a flame, it most definitely can't be used to duel another similar weapon. So how could we get closer to a more accurate lightsaber technology?

The Laser Sword

In *The Phantom Menace*, a young Anakin describes the lightsaber as a "laser sword."

Like a lightsaber, a laser can cut through metal and can cause great damage on contact. This is because they carry a great deal of energy. The lasers used to cut through metal can require hundreds to thousands of watts.

"Laser" stands for "Light Amplification by Stimulated Emission of Radiation." It essentially works by using energy to stimulate a substance into emitting photons. The aforementioned substance is known as the gain medium and can be made from a gas, liquid, or solid.

The first ever laser used a solid ruby crystal as its gain medium. This produced a red light beam. Other gain mediums can produce other colors, depending on the particular wavelength of the emitted photons. For example, larger wavelengths have lower energies and produce red light, while the smaller wavelengths have higher energies, producing blue light.

Other colors also exist, including green and yellow.

So the kyber crystal in the lightsaber could make sense as a possible gain medium. Then slight impurities in the crystals could give a lightsaber its characteristic color. This would also mean that different colored lightsabers have different energies.

However, as light waves are invisible, these beams could only be seen if they reflected off some surface or were scattered by particles in the surroundings. So a laser saber would actually be invisible, which isn't what happens in the movies.

This is not the only obstacle in the way of this method.

The Problems with Lasers

Everyday laser cutters, used for cutting sheet metal as thick as fifteen millimeters, can typically work at powers of five hundred to six thousand watts. The thicker the sheet of metal, the more power is needed to cut through it. It's been estimated that a lightsaber would need fifteen to twenty million watts of power to cut through metals like we see in Star Wars.

For comparison, Lockheed Martin needed a thirty-thousand-watt laser to disable a vehicle one mile away. And that device needed to be mounted on a truck. However, the most powerful laser ever built has a beam with a peak power of two thousand trillion watts (two petawatts); that said, the laser only operated for one trillionth of a second.

A major issue with laser beams is that, once they're emitted, they keep on going until they're absorbed or reflected. This means a laser saber would need some other mechanism to keep it contained within a fixed length. The laser beam would also suffer the same problem as the flamethrower saber in that it couldn't be used to block the strike of another weapon. One light beam would travel straight through the other.

However, in an online discussion between astrophysicist Neil Degrasse Tyson and particle physicist Brian Cox, it was suggested that at ultra high-energies there's a probability that gamma ray photons will actually bounce off each other. Tyson subsequently proposed that if you could make two gamma ray lightsabers, then you could possibly duel.

Don't get too happy yet; we're talking ludicrous amounts of energy here. Additionally, the power source for industrial lasers is usually mains power, while even the smaller high-power-laser-producing units are typically bigger than a bathtub. (Obviously not something you'd want to be carrying around.)

So if not a laser, then what?

The Plasma Saber

Plasma is regarded as the fourth state of matter after solid, liquid, and gas. It's what lightning and stars are formed from. In plasma, most of the electrons are separated from the nuclei of the atoms.

Similar to lasers, plasma can be used to cut through metal, too. The color of the plasma can also be affected by temperature. Higher temperatures, in the region of 14,700°C would give a blue plasma saber, while lower temperatures of about 727°C would result in a red plasma saber.

However, plasma that could cut through anything would have to be heated to about 7,000°C, so a red lightsaber might literally not cut it.

It would have to be confined in a suitable blade length and shape. Physicist and futurist Michio Kaku proposed the use of a telescopic ceramic core, which the plasma seeps out of, while being contained within a magnetic field. This would give the blade rigidity while keeping the plasma together.

Most lightsaber ideas make use of the charged nature of the particles in plasma, which allows the plasma to be confined within a magnetic field. This field would have to somehow be formed in the shape of a blade, and capped at the end, too. Any plasma escaping out of the end would reduce the blade's pressure and, thus, its heat intensity.

The Tokamak is a magnetic confinement device that holds plasma within a donut-shaped magnetic field. It's used to contain thermonuclear fusion reactions, which generate lots of heat. These are the reactions that happen in stars. The magnetic fields are necessary because no materials can withstand the heat of the plasma created in the fusion reactions.

The outward thermal pressure of the plasma must be matched by the magnetic pressure used to contain the plasma. The ratio between the two is called the plasma beta. Typically, the confining magnetic pressure can be two-and-a-half to ten times greater than the outward thermal pressure. For anything but a cooler red plasma saber, we would need a magnetic field stronger than that of the Large Hadron Collider at CERN.

Lightsaber Duel

What would happen if you dueled Darth Vader with such a weapon?

The magnetic fields of the weapons would interact. Depending on the direction of the field on each blade, and the direction of the blades relative to each other, they would either repel or attract. If they repel, then let battle commence. However, if they attract, you could be in for a serious problem.

According to Space Plasma Physicist Dr. Martin Archer, a light saber duel could be disastrous. When plasmas with different magnetic fields collide, a phenomenon called magnetic reconnection can occur. A huge amount of energy is released in the process, and searing hot plasma is ejected at high speed. Not good.

To avoid this, the combatants would have to ensure that the magnetic fields matched. This would depend on the angle between the magnetic fields (or blade angles), the plasma betas, as well as the distance between the blades; in other words, not the main things likely to be on your mind in the midst of a battle.

So, despite many solutions to making things that might look like lightsabers, the practical problems of developing an actual lightsaber are still vast. Even if we had the power, we'd still run the risk of killing our enemy and ourselves at the first strike.

WHAT WOULD IT BE LIKE TO CONTROL THE FORCE?

In the Star Wars galaxy, all life-forms are said to have midi-chlorians that allow sensitivity to the Force. Not everyone has the ability to use it to achieve superhuman feats, though.

In our own Galaxy, we haven't found anything like midi-chlorians. However, we do have species that have remarkable abilities. There are many things we humans can't do as well as other species, but something we do excel at is our brainpower. As such, the development and application of science and technology has probably been our strongest ally in achieving extraordinary feats.

In the case of the Jedi, they possess even-more-superior intelligence as well as bodies that can be pushed beyond normal limits. Even Yoda can dispense with his walking stick when it's time to rumble. This is because the Force is their strongest ally in achieving extraordinary feats.

The Force gives them the ability to control people's minds, see the future, and levitate objects, amongst other things.

What would it be like to have these abilities?

"He Can See Things Before They Happen"

Jedi have the ability to sense future events through the force. It's known as "precognition."

Precognition implies that information from a future event can be perceived before it has technically been sent, which goes against the principle of causality, i.e., cause and effect.

Yet still, we can actually predict future events without violating causality.

In the UK, the old adage "red sky at night, shepherd's delight . . ." helps a person approximate whether the next day will be sunny or not.

Modern weather forecasting is just a more sophisticated version of this saying, where personally observing a red sky in the evening has been superseded by more accurate scientific methods such as satellite observations of clouds. However, the further forward we try to predict, the less accurate the prediction will be.

"Difficult to see. Always in motion is the future," says Yoda.

While it could be useful to see the future, it also gives rise to problems of determinism, i.e., is the future fixed and already determined, or can we exercise our free will to change it?

The main thing is that there will be some future events that are easy to affect. For example, Lando could have folded his bet with Han Solo if he foresaw losing the *Millennium Falcon*. On the contrary, there are other future events that can't be affected, like Yoda's material form reaching its twilight and dying off naturally.

Even if you could see the future, you could only do so much with it.

Telekinesis

This is the ability to move things without an apparent physical means.

It's a Jedi's knowledge of the Force that allows them to perform their "mystical" tasks. Knowing how to affectively tap into and apply the Force can allow even the biggest objects to be moved by the smallest of Force users.

In the third century BC, the ancient scientist Archimedes said, "Give me a place to stand and I will move the whole world."

He was referring to the fact that, with the right application of leverage, it's possible for a normal person to apply forces that exceed their own strength. The lever provides a mechanical advantage where a small input gets a bigger output. Levers, along with devices like pulleys and gears, are literally used to manipulate force.

Visualize Luke on Dagobah wondering how to get his X-wing out of the swamp. While he and Yoda are meditating on the matter, Archimedes emerges out of the woods. He applies some cleverly positioned ropes and pulleys, then winches out the X-wing with the help of a swamp dragonsnake.

So, to move objects more massive than him, it's possible that Yoda manipulates the Force to act like a lever or pulley system. In this way, his training allows him to manipulate objects by applying the Force, whereas we have to be more hands-on by developing and utilizing mechanisms.

"I Don't Wanna Sell You Death Sticks"

Picture yourself at the bank and you need to make a withdrawal.

The old Jedi mind trick is possibly the Force technique with the most potential for selfish or evil acts. It involves influencing someone's mind and is usually done for personal gain.

With a wave of a hand and a few suggestive words, a susceptible person seems to enter into a hypnotic trance, repeating the suggestion back to you to indicate their compliance.

This Force ability seems very much like hypnosis, which is big business nowadays. It's used for therapy as well as entertainment. The main difference is that in these cases the hypnotist has the person's permission to be hypnotized. The Jedi don't.

There are more subtle ways to influence people's minds, though. We're surrounded by advertising campaigns built to manipulate us into buying into some idea or product. Often they hijack our natural responses to things by appealing to our emotions or including sexual imagery.

The Jedi basically use the ability to force their desires upon a person's mind, and they're meant to be the good guys! Imagine the chaos if that skill made its way out into society.

"Fear Is the Path to the Dark Side"

Good against evil is the conflict at the heart of Star Wars. This is particularly evident in regard to the Force, which has a dark and a light side.

Jedi use the light side of the Force for knowledge and defense, observing the Jedi code to steer them away from the corruption of the dark side.

However, the ancient enemies of the Jedi order, the Sith, were followers of the dark side of the Force. Their ultimate goal was to rule the galaxy. Both the Jedi and the Sith sought to develop their knowledge of the Force, but the Sith were prepared to take it to places that the Jedi code won't allow.

The seductive possibilities offered by mastering the dark side led to the gradual transformation of Anakin Skywalker into the Sith Lord Darth Vader. Anakin had fallen into the same trap as the medieval character Faustus, who bargained away his soul to the devil in exchange for knowledge and power.

Science fiction often portrays evil scientists who go a step too far in their pursuit of knowledge or power. This type of situation is mirrored in science, too, where some scientists push the boundaries of ethical practice in the pursuit of knowledge—some even stepping way over the line into the unethical dark side of science.

Around the time of World War II, the Japanese Unit 731 infected and experimented on humans. They were responsible for hundreds of thousands of deaths. There were also Nazi scientists, such as Joseph Mengele, who were considered evil, having also experimented on humans.

The development of weapons that caused horrific injuries could also be considered as a dark side of science. These include mustard gas, gunpowder, and the nuclear bomb. These technologies were all implemented in times of war, which seemed at the time to justify the darker side of their use.

With science always trying to push the boundaries of knowledge, there'll always be a pull to the darker side for some.

Force Lightning

The evil Sith Lords emit blinding bolts of lightning from their hands to punish whoever is unfortunate enough to be on the receiving end. Only the most powerful followers of the dark side were able to conjure this ability.

The Sith aren't the only living things that have the ability to shock with electricity. In fact, animals in our own world possess this ability. The electric eel (which is actually not an eel, but a knifefish) can emit a six-hundred-volt pulse of energy to stun its prey. It uses special electric organs in its body to produce the electricity. This is known as bioelectrogenesis.

The electric eel uses its high-voltage shocking ability for both attack and defense. The Sith appear to do the same, although they wield their skill with a fiendish pleasure.

We don't need to wonder what humans would do if they had the ability to project bolts of electricity. We've developed fifty-thousand-volt tasers that can deliver twelve hundred volts of electricity down a ten meter wire. Fortunately for assailants, the low-amplitude current of tasers protects them from being fatally shocked.

The Sith don't use wires, though, which means they need a lot more voltage to push the electricity through the air. Rhys Phillips works with the Lightning & Electrostatics team for Airbus Group Innovations. He says, "Assuming this is in air, which has a breakdown voltage of three million volts per meter—you'd need thirty million volts to jump ten meters."

So, if we had this ability, we'd have to raise ourselves thirty million volts above our opponent. However, the spark could just arc to the ground before reaching our enemy, so we might need to shoot laser bursts first to ionize the air. This creates what's called a laser induced plasma channel, which the electricity can travel down.

In conclusion, we may not have access to the Force, but we have our own ways of manipulating nature to achieve our goals. It could be said that science and technology is our Force, and in that light we already have an idea of what it's like to control it. Who knows what it will enable us to achieve in the coming years.

HOW COULD KYLO REN STOP A BLASTER SHOT IN MID-AIR?

Put yourself in Kylo Ren's position. Your troops have decimated the village and rounded up its remaining inhabitants.

You emerge from your ship to question Lor San Tekka, surrounded by your troops and the villagers. You have a simple request . . . the map to Luke Skywalker. Instead of complying San Tekka openly defies you, winding you up with talk of your family history.

But you're the great Kylo Ren. You need to make an example of him. So of course . . . you strike him down with your crossguard lightsaber.

Immediately you sense an imminent threat in the form of an energy blast heading your way . . . so what do you do?

A solution becomes obvious, being a Force user with a lightsaber in your hand. You parry the blaster shot!

On second thought, any old Jedi could do that. So instead you use your extraordinary reactions to dodge it and Force choke whoever was stupid enough to mess with the mighty Ren! You re-examine and determine that that's far too Vader.

Finally, you come to a realization: How better to showcase your superior grasp of the Force than to stop both the blaster bolt and aggressor in one fell swoop?

You swing around, raising your hand. Instantly the bolt and the aggressor, Poe Dameron, become your puppets.

Now while the blue bolt of energy is left crackling and vibrating in mid-air, let's take a quick moment to reflect on the situation.

Kylo Ren

Kylo is the troubled grandson of Darth Vader. Although the Force runs strong in his family, it's clear he has some major insecurities and daddy issues.

Thirty years after Vader's death, Kylo Ren has stepped up to try and fill his boots. He worships his grandfather's memory, often talking to the disfigured remnants of his helmet. He wants to embrace the power of the dark side, and intends to unleash it on everybody else.

This tantrum-prone master of the Knights of Ren is a firm follower of the dark side, but he also took lessons from the light. Despite earning the nickname "Jedi Killer" after being tasked with hunting down remaining Jedi for the First Order, Ren becomes conflicted when he comes up against Rey.

His unique knowledge of the Force allows him to conjure up tricks that are rare to both the Sith and the Jedi. The most visually impressive of these was the ability to stop a blaster bolt in mid-air.

This feat requires some major action at a distance on his part. He's gone one step further than Vader—or any other Sith or Jedi—to be capable of affecting a blaster bolt without being in close proximity to it. Furthermore, the bolt carries on along its trajectory once he leaves the vicinity. This could be because of the way he used the Force to stop it, or it's possible that he decided to send it along its path before leaving.

The question arises: How could he stop the blaster bolt in mid-air?

Stopping a Blaster Bolt

Blasters are said to deliver bolts of "intense plasma energy" that impact as a searing concussive blast. They do this by converting gas into a glowing particle beam capable of melting through targets.

There are many different sources of plasma in our universe, including lightning, stars, and neon lights. They can differ in temperature and in how densely packed the electrons are within them.

Like a gas, the plasma will try to fill whatever space it's put in. So for the plasma to be sent through the air like a bolt, something would have to be fired that can carry the magnetic field that confines the plasma.

However, according to space plasma physicist & broadcaster Dr. Martin Archer, you wouldn't have to actually fire a special magnetic field source.

"Plasmas are incredible electrical conductors, so if you could induce a so-called 'bootstrap current' into the plasma bolt there's the idea that this induced current's associated magnetic field could (for a time) confine the plasma."

At the University of Missouri, scientists have managed to launch a ring of plasma as far as two feet in open air without any need for containment. The plasma reaches a temperature hotter than the surface of the Sun. It forms its own self-magnetic field, which holds it together, although it only lasts for a few milliseconds. It's also not very dense plasma, so although its temperature is high, not much heat is actually transferred to its surroundings.

So if Ren wants to stop the plasma charge in mid-air he could create a field to contain the bolt. Then it would just be a case of holding the field in place in the same way that he holds Poe in place.

Another interesting idea on how Ren could've stopped the bolt would be to warp space and time.

Messing with Time

According to general relativity, the more mass (and hence gravity) that exists in a region of space, the slower time will tick in comparison to a region where the gravity is lower. So if Kylo Ren were able to increase the mass enough in the bolt's region, then time would move slower for the bolt, relative to someone further away.

Let's conceptualize: Imagine you're Poe being dragged passed this bolt and eyeing it from one meter away. We're going to slow down time so much that in the ninety seconds that the bolt is held, it only moves by ten centimeters. This means from the bolt's perspective, a miniscule amount of time has passed (0.3 billionths of a second). How much mass would Ren need to give to the bolt to make time dilate in this way?

The answer: more than the entire mass of Saturn! Jakku, we have a problem.

The gravitational pull would be so great that even if someone were on the opposite side of Jakku, they would suddenly gain two tons of weight. For Poe and everyone in the vicinity, it's over straight away. That amount of mass, packed into a region smaller than the inner sphere of a Zorb, would collapse into a black hole. In fact, the whole planet would be swallowed up!

Okay, so that idea is blown out the water. Let's modify the rules slightly.

Let's assume the blaster bolt is really a short burst of light (which it's not), made visible by a smoky atmosphere (okay, plausible), and traveling at lightspeed. Would Kylo Ren be able to stop it then?

Messing with Light

The speed of light in vacuum is about 186,000 miles per second, represented by the letter "c." This is its unimpeded speed, and as far as we know nothing can go faster than this cosmic speed limit.

Even though the speed of light in vacuum is always a constant rate, as represented by c, light can actually move slower than that. This occurs when it's not traveling through a vacuum, i.e., through matter.

The amount a material can slow down light is called its refractive index. When light travels from air into glass, it slows down because glass has a higher refractive index. However, when the light exits the glass and goes back into air, it carries on at its original speed. Sound familiar yet?

This effect can be explained by how the light moves through different materials.

When there's empty space, light travels at its cosmic speed limit, c. When there are obstacles, such as the atoms in glass, the photons of light are absorbed and then re-emitted due to the electrons in the atoms. Similar to a person trying to walk through a crowded room, these obstacles cause the photons to take a little longer to get through the space than usual.

From this, we can see how the speed of light doesn't actually ever change, but it does take longer to get through different regions of space.

Now that we know light can seemingly be slowed down; how about stopping it altogether?

It may seem like a wacky proposal, but it has actually been achieved in some laboratories.

Lene Hau of Harvard University used a super cold (almost absolute zero; the coldest possible temperature) Bose-Einstein condensate to slow down a light pulse to the speed of a bicycle. Then two years later, Ron Walsworth of the Harvard Smithsonian managed to stop light altogether.

The light pulse begins at about one kilometer in length. They shoot the super cold gas cloud with another laser, which makes the cloud go transparent. This allows the light pulse to enter the cloud. When they turn off the other laser, the light pulse gets trapped in the cloud. In the process, the one-kilometer light pulse compresses to a length of about half the thickness of a human hair. When they're ready (after a minute in this case), they then shoot the second laser again to release the light pulse. As it leaves the cloud, it continues at its normal speed and length.

A Bolt from the Blue

So we're back behind the mask.

Poe Dameron's been escorted to your ship for interrogating and you're feeling like the all-powerful Sith lord you've always dreamt of being.

You've been holding the blue bolt for almost ninety seconds now and you didn't have to destroy the whole planet to do it. Oh, how grandfather would have been pleased.

As you leave you hear curse words being thought in your direction. You turn to face this insignificant pawn. FN-2187.

You'll deal with him later.

You head back to your ship, keen to interrogate Poe, but one last trick to cap your megalomaniacal display. Instead of releasing the frozen blast, knowing the plasma would dissipate, you propel it on its original path, just to show the power of the dark side.

IS DARTH VADER TECHNOLOGY POSSIBLE?

He's the menacing figure who appears through the smoke, heaving deep, steady breaths through his facemask.

Dressed head-to-toe in all black and adorned in a cape and Stahlhelm-like helmet, his very presence breeds fear. He was even booed by some cinemagoers when he first appeared.

According to Ralph McQuarrie, an artist who was brought in to paint a few scenes from the script to persuade investors, Vader was originally conceived as "a tall, grim-looking general," who had the ability to cross between two ships in space. For a feat like that, McQuarrie argued, he would need a suitable outfit including some kind of method to breathe in space, thus the idea for the infamous breathing apparatus was born. The descriptions in the early manuscripts are the primary reason Vader's iconic features were invented.

However, space travel did not become the eventual reason for Vader's suit. George Lucas would later write in the backstory of Vader's fall into a volcanic pit on Mustafar, revealing that Vader's iconic look was a result of his injuries.

This brings us to the memorable scene in *Revenge of the Sith.*

The Suit

After Vader's perilous descent toward the river of lava and his clothes igniting, he suffered drastic damage to his skin.

Burns victims are often given skintight compression garments that can reduce the extent of particular types of scarring. Even though they would not be intended as a permanent solution, it's very possible that Darth's suit may have served a similar function at first due to his serious burns.

Vader's hermetically sealed suit has ten protective layers and could apparently even allow him to survive in space. To provide the same protection as a real space suit, it needs the ability to withstand the vacuum of space, an effective cooling system, an oxygen supply, and armor layers to offer a level of protection against things like micrometeoroid impacts.

Real space suits such as the Extravehicular Mobility Unit (EMU) can have as much as fifteen layers and are much thicker. Layers include an inner liquid cooling garment, a pressure garment, a thermal micrometeoroid garment, and an outer layer. (Oh, and don't forget the Maximum Absorbency Garment; yes, it's pretty much an adult diaper.)

The size of the EMU means there isn't much maneuverability, so a thin Vader-style suit would be much more appropriate.

Professor Dava Newman of Massachusetts Institute of Technology (MIT) has been working on such a design, which has been dubbed the BioSuit. It would be skintight and elastic, behaving like a second skin while applying a compressive mechanical counter-pressure that would support an astronaut in the vacuum of space.

To achieve the counter-pressure, the suit would include coil actuators made from shape-memory alloys that can snap back to their original shape when heated. This would also make it easier to don and remove the skintight suit.

Man in the Mask

In fire victims, inhalation is one of the most common causes of death, and the upper respiratory tract bears the brunt of the injury, particularly the trachea. Subsequently, Vader needs his mask as part of a unit that aids his breathing and talking, leading to his infamous scuba breathing.

Air temperatures near lava flows can exceed 49°C, depending on atmospheric conditions. A team at Cornell University has looked into tracheal burning from hot air inhalation. They found that a person inhaling temperatures above 85°C for more than twenty seconds will sustain tracheal tissue damage. They also say that lower-temperature damage is possible when the exposure time is increased.

In a 2014 paper, doctors Ronni Plovsing and Ronan Berg looked into the possibilities surrounding Darth Vader's lung injuries. They suggested that he may be "an example of acute and chronic respiratory failure following severe burns and thermal lung injury."

Based on his hermetically sealing helmet and the nature of his breathing, they propose that his mask may be acting as an advanced bilevel positive airway pressure (BPAP) system.

A BPAP machine provides a variable inward pressure to make breathing in easier and a lower outward pressure to help with breathing outward. The provider of the machine usually presets the breathing levels. Vader's, however, is probably automatic, if not controlled by the buttons on his suit.

However, while Darth Vader is pretty much always in his gear, according to Plovsing and Berg, lifelong BPAP treatment would be a rather unconventional option for today's medical practices.

More Machine Than Man

Some ailments require surgical removal of a person's larynx, called a laryngectomy. Even with the removal of their voice box, they can still talk with the help of an artificial speech aid known as an electrolarynx.

The older versions were monotone, and often described as robotic sounding, but newer versions have the ability to portray more of the tonal changes in speech.

As such, there's no reason that Darth Vader's voice would become deeper through a similar device, particularly when voice-changing software is readily available nowadays. We'll assume that it was done to make him sound more menacing.

A panel of buttons sits dead center on his chest, which is used to operate many of the implantable medical devices (IMDs) that keep him alive. Darth Vader relies on the suit to keep him topped up on a good supply of medicine, oxygen, and nourishment as well as supporting his everyday bodily functions.

He's not alone in having button-supported bodily functions, though. More and more people these days are having similar devices installed in their bodies.

In the UK, 6% of adults and 19% of children with type 1 diabetes use an insulin pump that stays connected to the body. This allows them to modify the amount of insulin they take by pushing a few buttons. The same applies to implanted gastric stimulators and pacemakers.

Technology is now becoming so advanced that a new breed of pacemakers can be implanted directly into the heart without surgery. Some, such as the WiSE CRT system, are even as small as a grain of rice.

Cyber Prosthetics

We've come a long way since peg legs and hooks, with an increasing use of clever electronics to enhance the experience of amputees. Real-life leg injuries like Darth Vader's require an above-knee prosthesis. This includes knee joint, ankle, and foot components—each of which possesses its own design difficulties.

The latest developments use microprocessor-controlled hydraulic knee and foot units. These units sense and adapt to a person's movements to offer them more natural motion and greater freedom. The wearer can also simply switch between different pre-programmed modes to allow tasks such as cycling or driving to be more easily accomplished.

When it comes to amputations of the arm, the prosthesis technology is even more intricate.

The bebionic3 is one of the most advanced. It contains powerful microprocessors that monitor and allow precise control of each finger. It can also automatically adjust the grip when it senses that a held item is slipping.

There are more advanced robotic hands out there, although they're not yet suitable for use as prostheses. Researchers from the University of Washington have produced what has been described as the most amazing biomimetic anthropomorphic robot hand. However, the complexity of getting it to function properly is in obtaining adequate control signals from the human to the hand.

Cyber Control

Darth Vader is connected directly to his electronics. His helmet apparently contains neural needles that pierce into his skull and spine to allow control of his cybernetic limbs.

Currently, to control prosthetic hands requires sensors that respond to tiny muscle twitches in the stump of the wearer. The sensors activate various pre-programmed hand positions. In the near future, similar to Darth Vader, prosthetic devices will be connected directly to an implant that is attached to the bone in the stump. The patient's nerves will be surgically connected to the implant, providing a link from the brain to the robotic arm.

US Defense Research Agency DARPA has supported the building of a robotic arm that links to the brain. Wires are connected directly to the motor cortex—the part of our brain that controls our muscles and the sensory cortex—where we receive sensory signals from our body. This way the brain can send and receive signals from the robotic limb in the same way it would a normal arm and hand.

The technology has already allowed a paralyzed man to manipulate objects and feel.

An emerging technology is neuroprosthetics, in which the brain is connected directly to a computer. Simple setups allow on-screen selections to be made, while intermediate units could control a wheelchair. However, the most complex arrangements would be used for controlling artificial limbs.

Considering the options available for human augmentation with electronic devices, we could soon be entering a cyborg age—if we haven't already.

In the 1960 paper that coined the word cyborg, the scientists argued that

> *"Altering man's bodily functions to meet the requirements of extra-terrestrial environments would be more logical than providing an earthly environment for him in space. . . . Artifact-organism systems which would extend man's unconscious, self-regulatory controls are one possibility."*

More than half a century on, we haven't used these technologies to augment humans for space travel, but their presence has been flourishing in the medical industry. It's just a matter of time before a genuine Darth Vader is walking the streets.

IS JEDI MIND CONTROL A POSSIBILITY?

So you've driven up to a venue and you're stopped by security; but you really need to get in.

What do you do? Well, you wave your hand and tell him that he must let you in. He promptly obliges and you go on your way.

It would be fantastic if this were really possible, but then again it really depends on who's got the ability. You'd hope that such tricks would only be available to people who intended to use it for good reasons.

Of course, what one person sees as a good cause might not seem so great from another person's perspective.

Besides the many ethical problems with its use, what are the chances of ever being able to manipulate a person's mind to cause them to do something against their will?

The Old Jedi Mind Trick

This is potentially the most invasive and unethical skill a Jedi has in their arsenal. On the other hand, it's also the one with the potential for providing the most entertainment.

Who wouldn't like the ability to get what they want from someone without putting in the usual effort of being nice to them, being patient, or offering some kind of trade-off? The Jedi mind trick is the immediate gratification of the Jedi whose agenda is clearly more important than the person they use the trick on.

It doesn't work on everyone, though. Jabba the Hutt reveals that he's no weak-minded fool when Luke tries out the old Jedi mind trick on him.

Toydarians like Watto are naturally unaffected, too, with money being Watto's major persuasion. People can also be trained in the ability to resist the Jedi mind trick. Clearly, the Force is a great ally, but it's no match for a disciplined or focused mind.

Mind control isn't limited to the Star Wars universe, though. It also doesn't require the use of the Force to be achieved. We're bombarded everyday with attempts to affect and even control our thinking via friends, strangers, and even family. Ever been persuaded to go on a trip you really didn't want to? Even scarier is that a lot of the time we may not even be aware when it's happening, i.e., being persuaded to buy something by the subtleties of advertising campaigns.

However, if we want to see some of the most extraordinary types of mind control, we just need to take a glimpse at the natural world.

Nature's Mind Control

In nature, there are many different mechanisms for controlling minds. Often it's a necessary part of how a species survives. Take insect communication, for example.

Ants use up to twenty different pheromones to communicate. The key thing is that they do not question the messages, they just respond to them; their responses are essentially programmed in. This enables a queen ant to use particular pheromones to control her workers.

The success of the colony requires that the ants are susceptible to being controlled. It's this type of mind control that allows the ant colony to survive. On the whole, their compliance is more of a strength than a weakness.

For some individual ants, though, compliance is at the heart of their impending doom. Weirdly, in the following case, it's not even another animal that's controlling them.

The spores of a fungus called *Cordyceps unilateralis* changes the way ants of the Camponotini tribe perceive pheromones. These changes cause the ants to do exactly what the fungi needs. Infected ants walk up to the underside of a leaf and just wait there until they die. The fungus subsequently grows out of the rotting ant's head to eventually send out spores to infect more ants.

This is interesting because it's an example of mind control across species, even though it only works on one particular tribe of ant. However, there are thousands of different types of *Cordyceps* fungi, and each specializes on a particular species of insect, whether it's spiders, ants, grasshoppers, or moths.

As mind control goes, this example is pretty gruesome, but its scope is very limited. The fungi do not choose the outcome; they merely create it as an evolutionary aspect of their survival. But what about more complex and varied forms of mind control, where a particular outcome may be chosen?

Compliance

When a Jedi makes a request of somebody, they are trying to get a compliant response. This is when somebody does something because they were asked to, even though they may have usually declined such an action.

The fact that Jedi mind tricks work on some and not others means that they're less like the pheromones or fungus spores that force an unavoidable and predictable response. The tricks are more like a form of persuasion where the Force can be used to help put the target into a compliant state of mind.

Major studies have been conducted on different types and levels of compliance where people have gone with something that is totally against what they would normally do.

In the 1950s, the Asch conformity experiments showed that if a person was in a group and the group went with a particular decision that they thought was wrong, they would tend to go with the majority of the group despite their own thoughts. This effect is increased when the "subject" is more uncertain about how they should respond, or if they perceive others in the group to be more knowledgeable than themselves.

In the 1960s, Stanley Milgram conducted experiments on obedience. People had to administer electric shocks at increasing levels, which they would do even though it made them uncomfortable. They continued because it was an authority figure that was instructing them. This implies that if a Jedi can exude authority, the person may be more likely to obey. According to Milgram, "Relatively few people have the resources needed to resist authority."

Another interesting study in 1973 was the Stanford Prison Experiment by Philip Zimbardo. Its aim was to see how people conform to roles of guard and prisoner when assigned the roles at random. The study had to be stopped early, but the participants had conformed strongly to their roles, acting in ways stereotypical of the role they had been given. Many took it too far. It showed that a situation can have more impact on our behavior than our natural disposition.

If a Jedi can give the impression of being an authority figure, there's no reason that somebody shouldn't respond to them. In the case of a stormtrooper, whose role is one of obedience to authority, it should be easier for a Jedi to affect their mind. This may be why stormtroopers seem so susceptible.

Compliance is a necessity for a smooth-running society. Sometimes compliance is for the good of society, but the Jedi mind trick seems purely for the whim of the Jedi. It's like being hypnotized against your will to do things you wouldn't otherwise do. However, in hypnosis the subject is usually a willing participant.

Hypnosis

Hypnosis is a technique where people enter into a trance-like state of mind in which they are more open to suggestion from somebody such as a hypnotist or Jedi.

Before hypnosis takes place, there is usually a period of preparation where somebody is helped to slip into the necessary state of mind. This is a practice commonly associated with the image of a person swinging a pocket watch in front of your eyes, saying "you are feeling very sleepy." This process typically takes a good few minutes at the least. However, the Jedi use the Force and a wave of their hand to instantly induce this state in someone.

Hypnosis doesn't work on everybody, though. A subject needs to have a degree of hypnotizability, which is based on how responsive to suggestion they are. A number of "hypnotic susceptibility scales" have been developed to measure this trait. We can imagine that stormtroopers would be ranked as highly suggestible on these scales, whereas characters like the Toydarians and Jabba the Hutt wouldn't even score.

But what about directly influencing or taking over a human's mind?

Mind-Reading Technology

Darth Vader and Luke Skywalker can sense the gist of each other's thoughts, to the extent that Vader even tells that Luke is thinking about his sister. In *The Force Awakens*, Kylo Ren takes mind control a step further to forcibly extract detailed information from people's minds, but it doesn't always work.

Although we can't yet interrogate people for information by scanning their brains, science is getting closer to understanding some of the processes behind our decision-making.

We assume decisions are made by our conscious mind. However, researchers in Germany have managed to detect the outcome of our decisions up to seven seconds before we're conscious of it ourselves. A next step for the researchers is to see if the unconscious decision can be reversed before the final, conscious decision is made.

In other research carried out at the University of Washington, a human brain-to-brain interface was made to let one person's thoughts move another person's hand.

The brain signals are read using Electroencephalography (EEG), which basically means it records electrical activity of your brain. It does this through a device worn on the head that resembles a swimming cap with connective wires sticking out of it.

On the other end, the respondent's brain is stimulated using Transcranial Magnetic Stimulation (TMS). This uses a coil placed directly over the part of the brain that they wish to stimulate, i.e., the region that controls the hand.

At the moment, the technology couldn't be used on a person without them knowing, because it requires a physical device to be worn. To achieve the same effect wirelessly poses huge problems in regard to the weak strength of the signals emanating from the brain.

However, it shows that information in our brain can be registered externally, even if in very rudimentary form. It also shows that we can influence the body using a signal sent from outside of it.

So mind control exists in many different forms—whether through suggestion and social pressure, technology, or a brain-controlling fungus. But as far as waving one's hand and making a suggestion, like with a Jedi mind trick, right now all bets are off.

✷ ✷ ✷